JOHN PONSONBY: RADIOASTRONOMER

tests and comparisons

Geoff Meggitt

John Ponsonby Radioastronomer

Published 2025 by Pitchpole Books

ISBN: 978-0-9575549-9-3

To Joyce

ACKNOWLEDGEMENTS

Numerous people contributed to this book by either providing advice or through interviews relating their experiences of John. I would like to thank them all.

In no particular order they are:

Ann Mathias, Jo Pauliny-Toft, Irma Kennaway, Peter Twyman, Jane Pine-Coffin, Maude Allen and Michael Gould for talks about John's private life. Marina Astrié for assistance with the French connection. Colin Hopper, particularly for helping me understand the maser issue. Dr John Churcher for insights into John's very broad interests. Professors Ian Browne and Peter Wilkinson at Jodrell Bank for an overview of John's career there and an appreciation of his contribution to radio astronomy.

The friends (now alas dispersed) he made at the boatyard in Northwich who enjoyed a cup of tea with him, entranced by his ancestry, connections and learning. His friends and others at the Manchester Cruising Association (now alas defunct) who thought him charming, erudite, arrogant or rude depending upon how well they knew him. I am grateful to Irma Kennaway for allowing me to use her sketch of John on the cover.

Finally, a very special thank you to Simon and Rosalind Lavington and Mac and Sue McLean, John's kindest and closest friends for so long. Without their encouragement this book would never have been started; without their enduring cooperation it would never have been finished.

Table of Contents

Index of Photographs

PREFACE

DEAN CLOSE

One of John's coats hung, empty and forlorn, on the post at the bottom of the stairs. On a whatnot close by the front door was a well-thumbed copy of Debrett's Peerage, ready at hand. Getting down the hallway was a squeeze between bookcases.

The main feature of one wall in the drawing room were long date-ordered sequences of journals: the Proceeding of the Royal Astronomical Society and those of the Institute of Navigation caught the eye. The shelf they were on hardly seemed strong enough to support such erudite weight but the dust that clung to the volumes suggested that the arrangement had been stable long-term. On the opposite wall was a larger more eclectic collection of books that reached floor to ceiling.

Otherwise the room looked like the store-room of an antique shop with uncertain furniture filling the space except for a small clearing in the middle – a viewpoint. Everything seemed to be behind something else and behind most things were elaborately-framed oil paintings of (you just knew) John's ancestors. There was an incongruous garnish of items: old VHF radios, a few sails, a bronze stern-tube for a small boat and a new-looking Loran-C receiver were scattered around. Every level surface had things on it.

Other rooms, nominally bedrooms, were disturbed seas of paper. The peaks surged around furniture and up the walls; the troughs revealed glimpses of dark carpet and were the pathways around the room. The rooms glistened with stacks of empty jam-jars in their hundreds awaiting those small treasureable items: nuts, bolts, nails, tacks, washers, resistors.

You knew one room was a kitchen because there was a sink. The draining board of this was covered in defunct hand-held VHF radios and GPS receivers with corroded batteries. Otherwise the usual trappings of kitchens seemed absent. However there turned out to be a cooker concealed behind a screen of jam-jars and a fridge we didn't, at that stage, feel inclined to open. In the larder we found a Seagull outboard motor lashed to a gas pipe.

In a small adjoining room a bright light flashed alternately red and green illuminating a bench strewn with electronic artefacts and tools. A soldering iron lay where John had put it down weeks, maybe months, before. Was there still a whiff of soldering lingering from then? Wire cutters, strippers and pliers lay about. Heat-shrink plastic tubes were heaped in a jumble. Power supplies hummed and winked, driving the flashing light. An oscilloscope looked on.

All round the walls, displays showed: our position in latitude and longitude, the weather forecast – streaming down a small green-glowing screen – and navigation warnings for different sea areas. You could see John on his own in the middle of the night, a curl of solder-smoke rising as he worked, data all around him.

It was all-in-all chaos on a scale I had never seen before and I was yet to see the garage.

The first obstacles were a massive metal-working lathe and several pairs of boat legs (yes, some boats have legs). Negotiating these required first a groin-threatening stretch and then a standing leap. All around was clutter: vacuum cleaners, a rubber dinghy, racks of electronic equipment, benches and shelves stacked with jam-jars now full of gleaming new nuts and bolts of all sizes and recovered brass wood-screws. The two, so-far inaccessible, tall things right at the back of the garage that I took to be part of a central-heating system turned out to be atomic clocks.

When I got back to the car and my wife (she had had a quick look around the house and retreated) I remarked that John seemed to have had a full and satisfying life. I'm not sure she fully agreed. She said it was more of a burrow than a home.

Over the months and years to come I came to see that she was right.

INTRODUCTION

My first encounter with John was in the late 1990s at a sailing event. We were both watching a Royal National Lifeboat Institution film of a lifeboat being launched down a slipway into very rough seas. I was quite moved by the scene and the knowledge that the crew were people putting themselves in harm's way for others.

I turned to the red-haired, shortish but sturdy, man standing next to me and said: " Makes the hairs on the back of your neck stand on end doesn't it". He paused for quite a while and then said, earnest and incredulous: "What on earth do you mean?"

I did my best to explain and he seemed to understand – although it was as though I had introduced him to emotional ideas that were quite new. I rather liked that and, indeed, him.

We were members of the same sailing and so met fairly often over the next twenty years or so. I enjoyed his company: his anecdotes, his knowledge of sailing, navigation and much else and his cleverness. I realised soon enough that not everyone else felt as I did. They saw him as arrogant, rude and pompous.

When he became ill, round about 2020, he realised that he would probably never sail his beloved *Vrelli II* again and he asked me to help him sell her. Before I managed to do that his condition became much more serious. I spoke to him on the phone a few times and then I heard he had died.

I was told in a phone-call from John's friend Mac who also mentioned that John had named me as an executor in his will. This was a complete surprise but I soon found out that he had also named someone truly competent, Rosalind Lavington, as an executor as well. Over the next several years (it was a complex affair) Rosalind did almost all the work and I just helped with the nautical matters such as the disposal of the boat.

However this was enough to enter John's world and realise quite how strange it was. I didn't need much encouragement to write a book about it.

I had John's substantial but patchy archive to work from and interviews (generally more like chats) with his closer friends. His quite numerous external publications and his PhD thesis were good sources for his technical career.

Explaining his career, with its successes and failures, has been a real problem: John's expertise was as an electronic engineer specialising in detecting and analysing the incredibly weak signals from radio telescopes. Just obtaining an image from these is a major challenge.

I have chosen to explain the trends in radio astronomy, while John was working, in the text and some of the things of what he did too. Two appendices go into a little more detail of masers (they were important in John's career) and the autocorrelator (which was quite novel and a major technical achievement). How well this has worked is for the reader to judge. If all he or she understands is that John was an extremely innovative and competent engineer in a ground-breaking field I may have succeeded.

My rather cryptic subtitle "..tests and comparisons" is from Michel Frayn's book *Towards the End of the Morning*. In it he claims that the modern equivalent of sin is failure – in tests and comparisons.

Finally a word about references. Where they are publicly available I have referenced them in footnotes or in an appendix but where they are to John's archive or to my interviews I have not. These documents will be deposited in some location yet to be decided and an electronic index will be provided.

1: ANCESTORS

John was proud of his ancestors, particularly the Ponsonbys; they were important to his story and maybe some of his attitudes. He was very aware of even his distant family (distant both in time and genetics) and I have interpreted that here as starting in the 1600s and running to around 1900, so to round about the birth of his grand-parents.

More recently, where he may have heard family tales of ancestors or where he might have known them himself and been influenced by them, I have gone into rather more detail.

EARLY PONSONBYS

A good point to start is with Sir John Ponsonby (1609-1678). He was a colonel in Oliver Cromwell's army in Ireland and did some campaigning from 1641 until the war was won in 1653. He soon became embroiled in the complications of the Act of Settlement of 1652 designed to recover some of the money that Cromwell had spent in trying to subjugate Ireland from his Catholic opponents.

Things changed rather with the Restoration of Charles II in 1660 when he became a member of the Irish Parliament and received a pardon from the King for his support for the Parliamentarian cause. A second Act of Settlement was passed in 1662 weakening the punitive effects of the 1652 one and the situation was further confused by the Act of Explanation of 1665.

He was granted some land in Ireland as part of the settlement and he settled down on the Kidalton Estate in Kilkenny. He died there in 1678.

He married twice. First to Dorothy Briscoe and then to Elizabeth Folliot – who had been married twice before – in Ireland. There was one child of the marriage to Dorothy (John born in 1634 in Cumberland) and three of the second Elizabeth, William and Henry all born in Ireland). The second marriage took place in 1658; daughter Elizabeth was born about ten years before that. He renamed Kildalton Castle "Bessborough" after his second wife.

THE ELIZABETH LINE

Our John was a descendent of the first marriage to Dorothy. But we will first follow the other, Elizabeth, branch.

Henry, the older brother, died before he could inherit so William became the 1st Lord Bessborough and Viscount Duncannon. He married Mary Moore and their son, Brabazon, became the 1st Earl of Bessborough. It was Brabazon's son William who dramatically changed the Ponsonby family's influence by marrying Lady Caroline. She was the eldest daughter of William Cavendish, the 3rd Duke of Devonshire and so seriously aristocratic.

Their son Frederick became the 3rd Earl and married the colourful Lady Henrietta Spencer. Frederick was physically abusive to her throughout most of the marriage; Henrietta's infidelities were numerous. When Frederick caught her *in flagrante*, coupling with one of her lovers, Richard Brinsley Sheridan the playwright, he demanded a divorce. A divorce was social suicide at the time and he relented and carried on gambling and abusing. She had two illegitimate children by another lover Granville Leveson-Gower, 1st Earl Granville before arranging his marriage to her niece.

Frederick and Henrietta had four children John William, Frederick Cavendish, Caroline and William. John William inherited the Bessborough title (though he preferred to be known as Viscount Duncannon) and became a Whig MP and served briefly as Home Secretary. Caroline married the Hon William Lamb and famously had affairs with the Duke of Wellington and Lord Byron (she coined the phrase "mad, bad, and dangerous to know" about him). After she died William Lamb succeeded to the title of Viscount Melbourne and became Prime Minister.

Frederick, the second son was to become Major-General Sir Frederick Cavendish Ponsonby GCMG KCB KCH. He served with distinction through the Peninsular wars and fought at the Battle of Waterloo (thus, as did his sister, serving under Wellington). He charged with his cavalry rather too far (possibly encouraged by the actions of Major-General William Ponsonby, a close relative) and was wounded in both arms and fell off his horse whereupon he was stabbed in the back by a French lancer. He survived this and when a French officer comforted him with some kind words and brandy he might have thought his ordeal was over. He was wrong: a French rifleman came and used his body as a shield to fire at the English. A little later he was ridden over by Prussian cavalry and roughed up by one of their countrymen searching for food. Finally a wounded British soldier crawled over his legs and died there. After waiting for hours he was taken by cart to Brussels to recover.

William was not so lucky. The brigade he commanded ignored the recall and reached the French guns only to find that their horses were exhausted. William's horse was trapped in mud and he was killed by a group of French lancers. The incident was captured in the 1970 film *Waterloo*.

In 1825 Frederick married Lady Emily Bathurst, daughter of the 3rd Earl Bathurst

His later career involved a command in Greece and he was for several years Governor of Malta.

He died suddenly in the Wellesley Arms, an inn near Basingstoke. He may have been aware of the irony: Arthur Wellesley was the Duke of Wellington whose arms his sister had notoriously been in. He and she had both served under Wellington.

There were six children from the Ponsonby-Bathurst marriage: Henry Frederick (b1825), Lt Col Arthur Edward Valette (b1827), Georgina Melita (b1829), Harriet Julia (b1830), Selina Barbara (b1835) and Frederick John (b1837).

The eldest son Henry Frederick joined the army as an Ensign in 1842 and served in several staff roles, including in Crimea, reaching the rank of lieutenant-colonel in 1855. He was appointed as an equerry to Albert, Prince Consort and in 1860 was promoted to colonel.

Henry Frederick married Mary Bulteel, one of Queen Victoria's

Maids of Honour in April 1861 and they had five children including John (b1866), Frederick Edward (b1867), Arthur Augustus (b1871), Magdelen (b1864) and Alberta Victoria (b1862).

After Prince Albert's death in 1861, he commanded a battalion of the Grenadier Guards in Canada, which had been sent there as a precaution during the American Civil War. I n 1868 he was promoted Major-General.

In 1870 he was appointed Keeper of the Privy Purse and Private Secretary to Queen Victoria. He occupied the office for 25 years. He was had a severe stroke in January 1895 and retired in May. He died in that November.

Several of the children of the marriage went on to have distinguished careers.

Son John was born in Windsor Castle and went to Eton. He was commissioned in the Sherwood Foresters but quickly transferred to the Coldstream Guards. He served in Uganda and South Africa. He fought in World War I as GOC of 40th and then 5th Division as a temporary brigadier and later as temporary Major General. After the war his major general rank was made permanent and he became GOC Madras District in India He retired in 1926 and died in 1928. He married Mary Robley in 1935 but had no children.

Arthur Augustus was born in Windsor Castle, was a Page of Honour to Queen Victoria, went to Eton and Balliol and joined the Diplomatic Service. He was a Liberal MP from 1908 until 1918. He strongly opposed Britain's involvement in World War 1 and became an active member of the Union of Democratic Control, a powerful and vocal anti-war pressure group. He contested Dunfermline in 1918 as an Independent Democrat and lost, probably because of his anti-war stance.

He joined the. Labour Party and won a Sheffield seat in 1922. He showed himself as a leader of the pacifist wing of the Labour Party claiming that the Great War was a conspiracy against ordinary people by profiteering capitalists. He popularised the phrase "the first casualty of war is truth" and set out to prove it with his book *Falsehood in Wartime*. He concluded that "international war is a monster born of hypocrisy fed on falsehood, fattened by humbug, kept alive by superstition , directed to the death and torture of millions."

In 1930 he was raised to the peerage as Lord Ponsonby of Shulbrede and continued his pacifist stance in the lead-up to World War II, opposing British re-armament and the guarantee to Poland (which he thought likely to lead to war rather than prevent it).

He eventually resigned from the Labour Party in 1940, opposing the wartime coalition led by Winston Churchill. He died in 1946.

Frederick Edward (known as "Fritz") had a good start: his godparents were the German Emperor Frderick III and Empress Victoria (which might explain his nickname). Eton was followed by a commission in the Grenadier Guards. He served in South Africa and, after promotion to brevet lieutenant -colonel, World War I. He made the first steps towards following his father in 1897 when he became Assistant Keeper of the Privy Purse and Assistant Private Secretary to first Queen Victoria, then to Edward VII and then to George V. He moved up to Keeper of the Privy Purse from 1914 to 1935 and he was Lieutenant Governor of Windsor Castle from 1928 to 1935.

Numerous honours were heaped upon him; when he retired in 1935 he was made Baron Sysonby of Wonersh. He died shortly after that.

THE DOROTHY LINE

The Dorothy line, that includes our John, starts with John Ponsonby, an only child, born to Dorothy Ponsonby in 1634. He married Anne Copley and they had eleven children. He died in 1703. We know that the eldest of these John was granted probate but not until 1724. John married Isabell Wood on 2 October 1705. One of the children was William born in 1684 and he seems to have had an offspring John born in 1706 and this John married a woman called Ann Millam. Their son, John, married Elizabeth Cropper and he became a solicitor in Whitehaven. Their son John is the starting point for

TREE1 in Appendix 1.

BUXTONS

Mary Catherine Buxton (John's grandmother) was the great-granddaughter of the anti-slavery campaigner Thomas Fowell Buxton MP. The Slave Trade Act of 1807 made the Atlantic slave trade illegal but slavery was still legal and prevalent in much of the British Empire. The abolitionist torch which led to the Act was carried in the House of Commons by William Wilberforce and, when Wilberforce stood down in 1825, by Buxton.

His main achievement was the passing of the Slavery Abolition Act of 1833 which made slavery illegal in much of the British Empire. For his work in the anti-slavery movement and the founding with Wilberforce and others, of the RSPCA, he was made a Baronet in 1840. He died in 1845.

Thomas Fowell founded something of a dynasty of public servants and social reformers. His son Edward North Buxton (the second Baronet), became an MP as did his other son Charles. Charles was a brewer and philanthropist. His third son Thomas Fowell became High Sheriff of Hertfordshire. Edward and Thomas followed their father in marrying a Gurney: Catherine, Rachel and Hannah respectively.

Edward North and Catherine had 11 children including Thomas Fowell (3rd Bt) who was briefly the MP for King's Lynn in Norfolk and became Governor of South Australia (for which he was awarded a high honour – a GCMG). Their son, Edward North was also an MP, an alpine climber and a noted conservationist.

Thomas married Lady Victoria Noel[1] and they had 13 children of whom 10 survived infancy. Son Thomas became the fourth Baronet, Charles Roden and Noel Edward both became MPs. Another, Harold Jocelyn , became a Bishop.

Charles Roden started out as a Liberal and won a seat for them in

1 The daughter of Charles Noel, the first Earl of Gainsborough

1910 only to lose it later in the year in another election. He switched allegiance to the Independent Labour Party in 1917. He stood for his new party several times in Accrington(where he won in 1922 and lost in 1923) and then in Elland (where he won in 1929 but lost in 1931 and 1935). His strengths were in his work behind the scenes. He married Dorothy Jebb.

Noel Edward was a Liberal MP for Whitby from 1905 to 1906 and then later for North Norfolk. He joined the Labour Party in 1919 and was elected as a Labour candidate for his old seat in 1922; he continued to represent it until 1930. He was briefly Minister of Agriculture and Fisheries in 1924 and again in 1929 until 1930 when he was made a peer and his wife Lucy succeeded him as the MP. He changed his surname and became Baron Noel-Buxton of Aylsham.

Charles and Noel were both concerned about the Balkans during the First World War and went on a diplomatic mission there to try to secure the neutrality of Bulgaria. While in Romania in October 1914 a Turk, Hasan Tahsin, tried to assassinate them. Both were wounded but recovered.

Harold Jocelyn took holy orders an eventually became the Right Reverend, the Bishop of Gibraltar. His title was a misnomer if ever there was one: his Diocese extended across Europe to everywhere except the United Kingdom.

One of the daughters of Thomas and Lady Victoria was Mary Catherine Buxton who was to marry Stewart Gordon Ponsonby and become, eventually, John's grandmother.

2: JOHN'S FAMILY

Stewart Gordon Ponsonby

Stewart Gordon Ponsonby was born in Bombay, on 18 October, 1856. His father was Captain Charles Ponsonby of the 7th Regiment of the Bombay Native Infantry and at the time DAQMG in the Seind Division ; his mother was Caroline Theophila (née Hutchinson) Ponsonby. His father died in 1858[2] and the family returned to the UK.

Gordon (he seemed to prefer to be called that) went to Fettes public school in 1870 where he excelled in sport becoming Captain of Football and of Cricket. He was the Challenge Clock winner for the best all-round sportsman in 1876.

He went up to Trinity College Cambridge in 1876 and was President of the Union in 1879. He graduated with a BA in 1880 and his MA came along in 1883. He was ordained as a Deacon (Durham, for Glasgow) in 1881, as a Priest (Durham) in 1882; he became Chaplain of the Iron Church, Moffat from 1881 until 1883. He was Chaplain of Trinity from 1883 to 1894 and Vicar of St Sepulchre's, Cambridge, 1885-94.

He married Mary Catherine Buxton, daughter of Sir Thomas Fowell Buxton, on 27 November 1889. Noel and Arthur were born in Cambridge. The family moved to Devon in 1894 when Gordon became Rector of Stoke Damerel and Rural Dean of Three Towns. They stayed there until 1922 and had their other children.

Gordon was Prebendary of Exeter Cathedral in 1907 and Chancellor in 1920 and Chaplain to the Bishop of Exeter from 1900

2 It was the time of the Indian Mutiny

to 1936.

He became Rector of St Mary-le-Bow, Cheapside in 1922 and died in post in 1938. While in London he was Chaplain to the , Drapers' Company from 1922-3; to the Grocers' Company, 1923, 1925 and 1932; to the Lord Mayor of London, 1923-4.

UNCLES AND AUNTS

Noel Edward Ponsonby

Noel was born in Cambridge on 14 January 1891 and boarded at St George's School, Windsor as a Chorister. He then went to Repton School and on to Trinity College, Oxford as an Organ scholar. After his Mus B he was organist at Royal Naval College, Dartmouth between 1912 and 1914 and then at Marlborough College until 1918. From 1919 to 1926 he was organist and choirmaster at Ely Cathedral.

It was at Ely that he met and, in May 1925, married the daughter of the Bishop of Ely, Mary Adela White-Thomson. The wedding in the Cathedral attracted throngs of people. The bride was dressed in "a sports suit, had a lace curtain as a substitute for the bridal train. Two tiny pages held the curtain under protest. In the centre of the curtain they placed their teddy bears.[3]"

Noel and Mary moved to Oxford in 1926 to take up a position as organist and choirmaster at Christ Church Cathedral in Oxford. It was here on 19 December 1926 that their son Robert Noel was born.

This was, in church music organist terms, a quite stellar career. He had even been shortlisted for the organist post at Westminster Abbey. A gifted choirmaster, he wrote church music and his *Five Fancies for Small Organ* is probably his best-known work. He was a strong supporter of a campaign for a School of English Church Music.

The Oxford Chronicle [4]described him in 1928:

"A certain quality of restraint in his manner prevented him from

3 The Daily Express, May 15,1925
4 The Oxford Chronicle, 14 December 1928

making many friends, but those who really knew him found in him a personality of much charm."

It also spoke of "a career of the richest promise". Tragically, this was in his obituary.

He was in good spirits at a Christ Church Senior Common Room dinner on the 10 December 1928 but he collapsed after he rose from the table. Assuming he had slipped, several people rushed to help him up but he was quite inert. They lifted him into a chair and within five minutes he was dead, certified by two doctors present. The cause of death was agreed as syncope – fainting and collapse caused by a sudden drop in blood pressure – a condition that is rarely fatal.

Noel was apparently in good health having recently passed an insurance medical with flying colours. There was no inquest as two doctors had been present to certify the cause of death which happened just a few days short of his son's second birthday. His funeral took place at Ely with a simultaneous Memorial service at Oxford.

Winfrid John

Winfrid was the birth name of St Boniface. The saint was born in Crediton in Devon in about 671 and spent some of his early life at a monastery close to where Exeter Cathedral now stands. He spent much of his time overseas, taking the Gospel to the Germans, and was martyred by local bandits in 754. In 2019 he was named Patron Saint of Devon by Devon County Council.

So our Winfrid had a name in keeping with his father's occupation (the Bishop of Chichester was a Winfrid too) but all too easily taken for Wilfred or Winifred. (I made the Winifred mistake and had him down as a daughter in the family tree for a start.)

We know rather little about Winfrid. He was born in January 1901 at the Rectory in Stoke Damarel and was there at the time of the 1911 census. His occupation was "school".

We know that he went off to Durham University in 1923 to study Arts.

On Tuesday 2 February 1926 the 20:38 stopping train from Sunderland to Durham was about half-a-mile from its destination when there was a bump suggesting that it had run over an obstruction on the line. The driver reported it at Durham before setting off back to Sunderland. A railway worker and a station porter were sent off to search the line. They found first a cap on the embankment and then, by torchlight, a headless body kneeling, still supported by its arms, at right angles to the track, feet pointing towards the embankment. The head lay a few feet away between the lines.[5]

There were footmarks in the field beside the track. It looked as if the person had climbed the boundary fence and then clambered down the steep embankment to the rails. An umbrella lay nearby.

Winfrid must have set out across the fields on this black, dark and foggy night, climbed the five-foot-high fence and set himself beside the line waiting for the rumble of a train. When its lights appeared from the fog he just leaned forward and rested his neck on the cold rail.

The inquest concluded that it could have been nothing other than suicide. His father the Rev Stewart Ponsonby, a witness at the inquest, reluctantly agreed. No-one could suggest a reason why this young man should take his own life. He was doing well in his studies and looking forward to his Finals, planning a future in teaching. He had seemed in good spirits when he left home in London on 21 January and had written a cheerful letter to his mother on the day of his death.

Asked if his son had always enjoyed good health his father replied:

"He was the least strong of my sons. If he had had to undergo four

5 "Suicide on the Railway: Durham Student's Sad Fate", Newcastle Evening Chronicle, 5 February 1926

years as a prisoner of war in Germany as another son had done, it would have killed him. He was a big boy, 6 feet 3, and his heart was not quite equal to his body."

Gilbert Jocelyn

Gilbert was born at Devonport on 4 May 1904. He went to Marlborough and then to Trinity College Cambridge where he read Geography and Economics from 1923 to 1926. He seems always to have had an interest in transport and his first job was with the Railway Research Service from 1927 to 1929 when he was offered a post as Assistant in Transport at the London School of Economics. He was appointed to a Lectureship there in 1934 and then, in 1950, to the Sir Ernest Cassel Readership in Commerce, specialising in transport. He retired in September 1964 and was elected an Honorary Fellow of the School in 1975.

His publications were not particularly numerous but quite influential. The first and perhaps most notable was *London's Passenger Transport Problem* published in 1932. It was followed by 15 journal publications including *The Problem of the Peak* in 1958 and in 1969 (so after retirement) by *Transport Policy Coordination through Competition.*

He was a member of the influential Greater London Group of academics – from a variety of disciplines – formed at LSE in 1958 with a remit to monitor, inform and influence policy in the city. The Group met weekly at LSE to listen to visiting speakers and debate matters. Gilbert continued as a member well into his retirement.

He married Sylvia Hollins at Freshwater on the Isle of Wight on 19 July 1930 after Banns were posted in his father's church , St Mary-le-Bow. His father officiated at what an (admittedly local) newspaper described as a "fashionable wedding on the Isle of Wight". Indeed, there were eleven bridesmaids and two pages and some 300 people attended the reception afterwards. The married couple left for their honeymoon in Scotland on the Fairbourne Ferry.

They had three children: Francis Noel, Edith Richenda (known to most as Richenda) and Stephen.

Gilbert died in Wimbledon on 18 July 1981.

Basil Leofric

While there are a few Leofrics in British history the most likely inspiration for the name was the one who became Bishop of Exeter in 1050. He was, according to some sources, born in Crediton and his remains are somewhere in the crypt of the Cathedral. So a local ecclesiastical hero.

Basil Leofric Ponsonby was born on 12 July 1909 probably at his father's rectory in Stoke Damarel. He was christened on 10 August 1909 in his father's church. One of the few enduring tokens of his life is a recently-auctioned christening mug:

> An art nouveau christening mug SILVER CHRISTENING MUG, CHESTER 1905
>
> Chester 1905, of flaring cylindrical form, embossed with a frieze of stylised plant forms, bears later inscription to base
> 'Basil Leofric Ponsonby from his godmother K.E.C. Aug. 10.
> 1909.', within fitted case.

We know little about his education other than that – following a family tradition – he graduated from Trinity College, Cambridge.

In August 1931 he travelled to Quebec on the *Empress of Britain* giving his occupation on embarkation as "Schoolmaster". When he arrived in Canada he said he was a "Student". He gave his Canadian address as Mr Vincent Massey, Batterwood House, Port Hope, Toronto. Massey was an extremely rich business man whose fortune built on the family tractor empire, a Liberal politician, diplomat and philanthropist. By 1931 he had already been in a War Cabinet and, from 1926 to 1930, he was the Canadian First Minister to the USA. In 1936 he was appointed High Commissioner to the UK and in 1952 he became Governor General of Canada. So not a bad sponsor for the young Basil.

We don't know how Basil spent his time in Canada but we do know that he went back in the following year and travelled across the country to Vancouver. He then took the venerable *SS Princess Marguerite* across to Seattle where he was admitted on 19 July as a 6' 2" fair, blond student who wore glasses. He planned to be in the States for a month and to travel to Los Angeles, Denver and Detroit.

He left the USA at Detroit on 8 September, presumably by steamer, and by 10 October was in lodgings in Toronto where he met up with a Canadian academic, Roy Daniells. We know this from Daniells's biography and we know from his archive that they had been writing to one another since May 1932.

Daniells was just starting work on his PhD at the University of Toronto and, as part of that, he was to go to London. This he did the following year arriving on 24 October 1933. He soon enrolled at King's College. He seems to have been unwell much of the time. A doctor he visited found nothing physically wrong but suggested a session with a psychiatrist because he thought that there was, in his own words, "probably a permanent kink in my mind due to religious fear and sex-repression." Whether the events of Boxing Day helped we will never know.

He had no doubt made contact with Basil soon after he arrived. Basil, learning that Roy had no plans for Christmas, arranged for him to be invited to a Boxing Day party at the house of Lord Noel-Buxton. He met the Liberal peer (Basil's uncle), the Bishop of Gibraltar, (Harold Buxton – another uncle) and Charles Roden Buxton (the Labour MP and yet another uncle).

Part of the conversation must have been about the assassination attempt Noel and Charles had survived while in Bucharest, Romania in 1918. Noel was hit in the jaw and Charles in the lung by bullets fired by a Turkish activist Hasan Tahsin. A road in Sofia, Bulgaria – Buxton Brothers Boulevard – still commemorates their visit to the Balkans.

Daniells must have been impressed by this high-achieving family and to meet one assassination subject would have been exciting but two at the same time was unbelievable. Enough to straighten the kink in his mind? It could have made it worse.

One of the things we learn from Daniell's biography is that Basil was in the employ of the Bishop of Gibraltar. The diocese was not restricted to the tiny island but extended to the Anglican communion across mainland Europe and the post was later re-named "The Bishop of Gibraltar in Europe" to reflect this. Quite what Basil's role was is not known. However, we do know that he was secretary of a religious society and that may have been the Student Christian Movement and his sponsor for that may have been his uncle Harold.

His father died on 11 July 1938, the day before Basil's 29th birthday. Basil attended his memorial service on 16 July and then on the 20th set out from Hull aboard the Finnish vessel *Arcturus* to take a party of church workers to Estonia. The route took them first to Copenhagen and then Finland.

On the night of the 21st at 11 pm, as the shhip approached Copenhagen, Basil was seen asleep in his berth by one of his companions.. When this man got up at 5 am the following morning Basil was gone. The alarm was raised at 10 am and a thorough search of the boat failed to find him; his passport and pocket-book were under his pillow. People remarked that he had been very quiet – some said silent – since leaving Hull.

The loss was reported to the police in Copenhagen and a consular enquiry was conducted when the ship reached Finland. The conclusion was that Basil had been lost at sea. The sea had been calm. Basil went deliberately onto the deck in the middle of the night. He had been quiet and possibly depressed. His father had died a few days previously. However, his brother Gilbert told the newspapers that "there is no question of Basil's taking his own life", commenting that he had been happy when he left to take the party to Estonia.

Is there a memorial to this sad young man other than a quite pretty art nouveau christening mug valued at £100-200?

Katherine Victoria

Katherine was born in 1896. and married an up-and-coming academic, Arthur David Ritchie, in 1921.

Arthur (born 1891) had been educated at Fettes College in Edinburgh and St Andrews, obtaining a degree in Chemistry. During the First World War he was a chemist in the Naval Airship Service. When the war ended he moved to Trinity College, Cambridge to do an MA.

It was there he married Katherine.

He won a fellowship of the College in 1922 on the basis of a thesis on the scientific method which was published as *Scientific Method* in 1923 and showed his growing interest in philosophy.

In 1925 he moved to Manchester as a Lecturer in Chemical Physiology and remained in that post until 1937, teaching and researching the chemistry behind the physiology of the body. His real interest, one suspects, lay elsewhere because his next book, based on lectures he had given at Trinity in 1935, was *The Natural History of the Mind.*

He was finally (aged 45) appointed to a vacant chair of philosophy at Manchester in 1937 to reflect his changed interests. He remained in Manchesert until 1945 when he became the Professor of Logic and Metaphysics at Edinburgh University. He was elected a Fellow of the Royal Society of Edinburgh in 1946. He retired in 1959.

After the move to Edinburgh he published books and articles on:

Civilization, science and religion, 1945
Science and politics, 1947
Essays in philosophy, and other pieces, 1948
Reflections on the philosophy of Sir Arthur Eddington 1948
British philosophers, 1950
George Berkeley's Siris, the philosophy of the great chain of being and the alchemical theory, 1954
Studies in the history and methods of the sciences, 1958

He died in 1967 shortly after completing a re-appraisal of George Berkeley.

His father, David George Ritchie (26 October 1853 – 3 February

1903), was a Scottish philosopher who had a distinguished university career at Edinburgh, Balliol College, Oxford and in 1894 was elected professor of logic and metaphysics at St Andrews.

The couple had two children, Justin and Clare. Katherine died in 1987.

Mary Felicity

Mary Felicity (she preferred to be known as Felicity) was born in 1898 and married the 4th Baronet of Hyderabad, Sir John Kennaway, in 19331.

The Kennaways originated in Scotland as clothiers but one branch moved to Devon to take advantage of the burgeoning wool trade. When the wool trade collapsed, because of the Napoleonic Wars, the family moved into the wine business but two brothers, John and Richard, decided that their future lay in India. Richard worked for the Board of Trade in Bengal and seems to have amassed a large fortune as a result. John became the first British Resident at the court of Ali Khan, Asaf Jah II, the Nazim of Hyderabad. The British Resident was nominally a diplomat but in fact was guiding and monitoring the Nizam as an exercise in indirect rule from London. So he had a major hand in the negotiations to establish a trading agreement between the Nazim and the East India Company. For this, in 1791, he was made a Baronet.

The brothers returned to the UK and, in 1794, bought the Escot estate in Devon and started a new life as country landowners. The original house was razed to the ground by fire in 1808 and was not rebuilt, by the second baronet (another John), until 1837. This John became High Sheriff of Devon.

When he died in 1873 the title passed to John Henry who trained as a barrister and became a Conservative MP for Devon constituencies from 1870 until 1910. He was made a Privy Counsellor in 1897 and appointed Companion of the Order of the Bath in 1902.

He was famous for his enormous beard, inspiring Edward Lear, a friend, to write:

There was an old man with a beard
Who said "It is just as I feared!
Two owls and a hen, four larks and a wren
Have all made their nests in my beard."

When at Escot he joined in with the hunting and shooting that were part of a country gentleman's life. He took care of the gardens, commanding a large team of gardeners for the purpose. He was also an enthusiastic, part-time soldier for 42 years, serving in the Rifle Volunteers as a commanding officer and then as an honorary Colonel.

At John Henry's death in 1919 the estate and title passed to his son John (sometimes called "Jack"). The estate had fallen into disrepair during the war and John started a major tree-planting exercise as well as a general mechanisation programme – needed because of labour shortages caused by the casualties of the war.

He married our John's aunt, Mary Felicity, in 1931. He was by then 53 years old; Felicity was in her early 30's. They had three children: John Lawrence, Richard Noel and Mary Joyce (known as "Marigold").

When John died in 1956 the title (the 5th Bt) passed to John Lawrence who had prepared himself (or maybe, been prepared) by going from Harrow to Wye College to study rural estate management. He struggled for almost 30 years to manage the estate while diversifying and coping with the death duties resulting from the passing of his father. Poor judgement and bad luck meant that the estate suffered a terrible decline and in 1984 he, with encouragement from the wider family, stepped aside to let his son John Michael take over.

John Michael was in his early 20s and with his wife Lucy transformed the estate business into the vibrant and viable business it is today.

COUSINS

Robert Noel

Robert Noel Ponsonby (b 19 Dec 1926) died aged 92 on the 3

November 2019.

He had visions early of a performance career after Eton and Trinity Oxford (where he took an MA in English literature) and National Service with the Scots Guards. He returned to Trinity briefly as an organ scholar from 1948 to 1950 but perhaps realising the challenge and his strengths when a job came up at Glyndebourne in 1951 as an assistant to Sir Ian Hunter who ran both the Sussex company and its satellite, the Edinburgh Festival. Robert gradually took over responsibility for the Festival – it was always a stretches of resources anyway – and it was no great surprise that he was offered the job as separate entity when Hunter resigned in 1955.

He had for a while wanted to improve the theatrical content of the Festival and aiming high he had hopes of Audrey Hepburn in 1956, sadly negotiations broke down although Christopher Plummer delivered a fine Henry V instead.

The following year was a triumph and a tragedy. Maria Callas, at the pinnacle of her powers, was to come and sing five performances of Bellini's *La sonnambula* at the King's Theatre. Well, Noel thought she was. Through some oversight she thought it was just four and the diva is always right (this diva was anyway). A hasty note was slipped into the programme indicating that the performance was to be given by Renata Scotto because Callas was unwell. Photos were soon discovered that proved that she had been partying in Venice or, seemingly simultaneously, storming out of a Princes Street hairdressers. Ponsonby remained unruffled. The lady never came to Scotland again but many others did: Joan Sutherland, Margot Fonteyn and Otto Klemperer.

The Fringe had grown alarmingly since its beginning in 1947 and it clearly represented an opportunity for the main Festival. Noel had a success with the booking of the singer and comedian Anna Russell who entertained with a mixture of serious singing and a variety of spoofs in 1956 so when negotiation broke down for the 1960 season he asked his assistant, John Bassett, to stage a late- night revue instead. John knew of a chap called Dudley Moore, another organ scholar and a jazz enthusiast who was having some success in cabaret. Moore recommended Alan Bennett. John also knew of

Jonathan Miller and Miller recommended a fellow called Peter Cook. Thus was born *Beyond the Fringe*, a satirical entertainment. After a chaotic dress rehearsal Noel asked a colleague "What have I done?" He had, with Bassett, only changed comedy forever making it less deferential and more anti-establishment.[6]

He left the Festival in 1962 and ,after a short generally unsatisfactory break at the Independent Television Authority, he became manager of the Scottish National Orchestra to great acclaim.

In 1972 he left to become the Controller of Music at the BBC and thus took on the Proms, perhaps his greatest achievement. He stayed with the BBC until 1986.

Not having had quite enough of music festivals he devoted a further two years to the Canterbury Festival. He had been appointed CBE in 1985 and after Canterbury continued to write obituaries for the Independent. He spent a good deal of his time after his retirement in the Lake District, walking and bird-watching. He was a keen collector of Scottish and English watercolours.

In 1957 he married Una Kenny and in 1977 Lesley Black. Both marriages were dissolved and he had no children.

Francis Noel

Francis was born on June 14 1933, the son of Gilbert Jocelyn Ponsonby (1904 - 1981) and Sylvia Hollins. They had three children: Francis, Richenda (who preferred the name to Edith) and Stephen.

Francis shone as a child, precocious and erudite, and was a page at King George's coronation (quite why is not clear but he had a Coronation Medal which he liked to show people ever after with the pretence that it had been awarded for secret intelligence work.) As for precocious, he wrote to the Times when he was 16 crediting the Government with "elementary principles of socialism", in their attempts to reform entry into Dartmouth, where he had recently arrived, but deploring the fact that so few candidates had applied to

6 Ponsonby, Robert, "Oral History of Glyndebourne Opera" - Classical Music | British Library - Sounds,

"join the finest service in the world." A thorough review of the situation was required by the Admiralty.

The letter caught the eye of the Leader of the Opposition, Sir Winston Churchill, who invited Ponsonby to lunch, unaware of the writer's youth. Francis, probably wisely, declined.

Educated at Eton and Dartmouth, he married Sally Cocup, the daughter of the Deputy Chaplain of the Fleet in 1961. They had four children: Nicholas, Jane, Emma and Simon Mark Noel.

An early posting was as *aide de camp* for the Governor of Malta in 1959 but he was selected for the submarine commanding officers course (known as the "Perisher") in 1962. It is a course that has been run from the early days of the submarine service. Originally the Periscope course, it became a byword for a kind of civilised gentlemanly brutality: if you flunked the course you would never see the inside of a submarine again. You would have to make your way in the surface fleet.

Francis presumably made it because he not only passed but was the executive officer for the nuclear Polaris submarine *Resolution* commissioning the vessel and coordinating some of the test firings of the missiles.

When the submarine *Renown*, another of the nuclear fleet, surfaced under the *MV Moyle* in February 1969 it broke a few bits and pieces on the coaster but hardly damaged the submarine. But the Perisher discipline was strong and the commanding officer was relieved of his command. Francis took over.

He always ensured that his submarines set off with a cellar of good port and gave instruction to any officers who didn't know how to deal with the equally good Stilton that accompanied it. He didn't hold back when a dispute arose in the *Times* about the difference between shepherd's pie and cottage pie.

He was, as his obituary[7] notes, "good company, laughing loudly and first at his own jokes."

Renown was followed by a period in command of a frigate, *HMS Salisbury*, his final seagoing posting. The Third Cod War occupied

7 The Daily Telegraph 2003

him in the 1970s and he was Naval Attachee to Norway. His final appointment was as Assistant Director of Public Relations (Royal Navy). Typically it ended with a flourish. When the Falklands war broke out, he was in the South Atlantic making a publicity film about the ice patrol ship *Endurance*. He promptly volunteered to become chief of staff to the *Endurance's* commander, helping to plan the sinking of a submarine and the operation which drove off the attempted landing be the Argentinians on South Georgia.

After retirement he became a respected correspondent for the *The Officer*. He was choirmaster and chairman of the Petersfield Area Churches Together. He collapsed and died at a submarine officers' reunion on the 24 May 2003.

Rosie Swale

Rosie Swale is a descendent of the first-cousin marriage of Sophie Elizabeth Ponsonby and Justinian Gordon Ponsonby. One of their children, Cara Ponsonby, married Peter Griffin. Their son, Peter Ronald Ponsonby-Griffin became a Major in the Royal Engineers and married a Swiss woman. Rosie was born in Davos, Switzerland on the 2nd of October 1946. Her mother died when Rosie was just two years old (probably of tuberculosis) and she was sent to stay with her (by now bed-ridden) grandmother Cara in County Limerick, Ireland. After rather rudimentary home-schooling (she spent a lot of time in the countryside and rode her horse *Columbine*) she was sent away at age 13 to a boarding school in Cork.

She worked for a local newspaper for a while but quite soon the spirit of adventure beckoned and she hitch-hiked to India, Nepal and Russia. It rather set the pattern for her restless future life.

In her early twenties she met and married Colin Swale. They had a daughter, Eve, in 1969. Within a couple of years they bought a boat – a 30ft catamaran – and set off for Italy where Rosie gave birth to her son James.

This done, they set off, with Eve, for Gibraltar with a plan to sail to Australia. They left Gibraltar in December 1971, crossed the Atlantic, went through the Panama Canal, crossed the Pacific via the South Sea Islands and made it to Australia in 1973.

They came back via Cape Horn which made them the first to do it in a catamaran.

The adventures were well-reported by the tabloids – not least because she spent much of the time in the Tropics naked except for a pair of trainers. She was photographed in the rigging of the catamaran.

It led to a brief career as a glamour model and a notorious session for *Fiesta* magazine.

In 1983, now divorced from Colin, she decided to make a solo crossing of the Atlantic and become just the fourth woman to do so. She met her second husband, Clive Pope, as she was preparing her boat for the attempt.

She set out from Pembroke in Wales on 13 July 1983 and went via the Azores and Caribbean to arrive at Staten Island, across the bay from Manhattan, after 70 days.

It was here that John enters the story.

He had followed Rosie's adventures, she was after all his cousin, closely and had been in touch with her.

He happened to be in New York a short time after she arrived and decided he would give her a surprise by turning up unannounced at the Staten Island marina. He took the ferry from Manhattan to Staten Island and asked where her boat was. The security guard told him that the boat was indeed there and gave him the berth number but told him that she wasn't there – and wouldn't be back for a few weeks. John was obviously dismayed and the guard gave him the key to the boat for him to have a look. Later on the guard suggested (maybe it was getting dark) that John might like to spend the night on her. So he did.

What prompted the guard to extend this act of improbable kindness to an unknown stray Englishman John couldn't fathom.

The best he could do was suggest that it was his accent.

Rosie was to go on many adventures after that: long distance running and walking, crossing America, Rumania, Iceland, Albania,

Cuba, Nepal, Russia from west to east and Canada.

For some of these later runs she towed a tiny trailer with her provisions which served as sleeping accommodation at nights.

John stayed in touch with her until his death and, unsurprisingly, he seems to have helped fund some of her many projects.

The Franco-German Connection

Illustration 1: French Cousins (L-R) Robert and , Marina Astrié ; Pierre and Jacques Caudé

The following in a lightly-edited version of an account of the French/German side of the family written for John's memorial event by his cousin, Marina Astrié:

> The first time I met John was in spring 1939. He was in his pram in Hyde Park crying himself crimson !
>
> The second time it was in the summer 1947. In between it was war in Europe and during that period we got only few

black-and-white photographs from John and his mother, my aunt Jacqueline. Then I met him at last, a fat little boy of eight years old with ginger gold hair, you could not miss him...

Since then our friendship has lasted forever.

During all those years we met now and then, once in Escot (1950) at aunt Felicity's home where the family lived during the war, lots of times in Thame, in Paris, in the French Alps, on his trips around Europe or around the world. The last time we met was in 2019 at the wedding of Benoit Astrié, one of his favourites among my 13 grandchildren. He felt at home everywhere. No wonder.

As for Jacqueline, she was born in Germany in 1905 in a city now called Krefeld[8], on the left side of river Rhein, in Rheinland. That area had been a French Department once, under the administration of Napoleon,and for several years (from 1802 to 1815).

This town was known as "die Seidenstadt" (the city of silk), like Lyons in France, because of the industry of silk imported by the French Huguenots in the 17th century.

Jacqueline's father Carl and her grandfather, Carl the first – who was known as a poet – were important citizens in Crefeld, their ancestors being probably Huguenot immigrants who became wealthy with the time (but they lost everything in the last century due to the recession between WW1 and WW2).

Carl was fond of travelling and he would travel around Europe, from Scotland to Constantinople, across Paris and Switzerland. He used to take the Orient Express, at Victoria "four thirty" from Paris, to Genève, Simplon, Venice or Vienna.

In Vevey by the lac de Genève he met his wife Ursule and

8 Before 1925 it was called "Crefeld"

married her in Versailles in 1900. The young couple went back to Krefeld, to his current home which was the most splendid building at the corner of Rheinstrassse in Ostwall (still called the Kirdorf Haus – although it had been destroyed and rebuilt after the last world war).

Carl's lovely young wife, Ursule de Puthod, was the daughter of Comte Edouard[9] de Puthod, from old Noblesse de Robe, (since before the French Revolution).

Among their ancestors, one had married Blanche du Terrail whose ancestor was the famous 16th century Chevalier de Bayard Pierre Terrail.

During Napoleon's time, Jacques de Puthod, Edouard's uncle, was made general at the battle of Eylau. His name is printed on the Arc de Triomphe. Edouard himself, during the 1870 war against Prussia, left Paris in a hot-air balloon with important documents belonging to the Imperial government, Paris being assieged by the Prussian army.

After the war in 1872, Edouard married Amélie Reinhart, who was born in Colmar, Alsace which then became German until 1918.

No wonder that one of their 3 children (Ursule) was friendly with franco-german people from the left side of the river Rhein...

So Ursule (usually called Loulou), their second child, followed Carl to Krefeld in 1900 after being wed. The 70 war was far away, 30 years had passed.

This nice young couple then began their happy life in the (original) Kirdorf Haus at the best corner of the city. Three daughters were born : Elisabeth, Jacqueline – John's mother – and Claude (her father Carl wanted to name her "Zepeline", because it was the first time that this new airship appeared in Krefeld's sky in 1908). Claude, usually called Coco, (not Zepeline !), is my mother and mother of my two half-brothers Pierre (Peter) and Jacques Caudé.

The happy Kirdorf family moved later on into a three-

9 John was named Edward after him

storied house and garden on the Westwall (there were four walls built over the former medieval fortification : Westwall, Sudwall, Nordwall, Ostwall, long before the first world war, with footpaths I remember in the middle, patches of lawn and fountain called Springbrunnen)

It was an awful blow to my grandmother Loulou when the first world war started. My grandfather Carl was fortunately too old to go to war and fight against French people. Imagine, that on the French side, Loulou's brother Julien de Puthod was fighting in the trenches... She had also no news from her widowed mother Amélie and her sister Valentine who lived in Paris.

For four long years there was nothing to eat and heat their home in Krefeld. The three girls who were accustomed to spend holidays or even several months with their French grandmother at 60 rue de Bellechasse (7eme arr), almost forgot their French, but fortunatly not quite !

From 1918 to 1922 they could never visit Amama-Amélie de Puthod in Paris again. They could not go to her funeral as the border was closed, French soldiers were occupying Rheinland for some months.

The girls were grown up by the time the “années folles” (The Twenties) started. Crefeld’s society began to relive, exchanges began again with France and England.

Loulou used to give French lessons and speak French at tea parties to young ladies of prominent Crefelders (Do you know the young lady Ursula von der Leyen who is currently president of the European parliament in Brussel and who speaks French? Well, her ancestors probably attended my grandmother's French tea parties in Crefeld...as her family has been known in Crefeld since Napoleon when he came to Crefeld on a visit in 1802).

But soon a crisis followed. German people felt uneasy. The financial crash destroyed economy, called inflation...Hitler appeared as a hero and a saver... You know what happened then, less than 10 years after the first world war : War. Again. It was said after 1918 : ' never again'. Alas !

In the meantime everything had to be germanised: Crefeld became Krefeld and Carl, Karl. Nevertheless the family would spend time in Paris. Elisabeth studied Art at the Sorbonne and married a baron: Jean de Prez de Crassier.

As for Jacqueline, she went to Berlin to study and learn a new Italian way to educate children, called Montessori, but she also often came to Paris to attend the "Lundis de Valentine" in the Salon of her aunt.

This lively place attracted many young students from the United States and from England too. They were all learning French but Jacqueline learned English on the occasion. So did my mother Coco. She was the first one of the family to visit England and on her return she told her sister how marvellous it was there. (I never understood why she made that unhappy marriage in Krefeld).

Jacqueline was much wiser so she visited London and a certain Lady Ponsonby whose son used to go to the parties of her aunt Valentine in Paris and whom she had met...

This chap, Arthur, was now posted in Liberia, Africa, as vice-consul of Great Britain. Before he left he had told his hostess Mademoiselle Valentine how much he was fond of her sweet German niece! Arthur, who had been a student in Germany, was imprisoned in Heidelberg with other English students during the first war. Among them was the young future lord of Rausham, I do not remember his exact name, but his estate is not far from Thame. They had no bad memories of their captivity and they learned proper German.

Now the loop is tied. The beginning of a new story and the end of the old story: Jacqueline boarded a steamer to Africa, was married to Arthur by another English consul at the opposite side of Africa, (as Arthur could not do it for himself). They came happily back in time for the birth of John ...and war, again. I am sorry to tell.

For our Grandmother Loulou, (we children called her

Amama now), it was another blow: for a second time she had to part from her French kin.

Jacqueline, after saying good bye to her in Krefeld, was far away in Africa, married to Arthur. Elisabeth, who happened to be in Krefeld at the time, took the last train to Paris: I remember seeing the last carriage disappearing from the empty station of Krefeld in 1940.

But the worst was to come. My mother and I now lived with Amama in her home at 136 Westwall. There was again nothing to eat. They had to queue in turn to get some food on the market. I nearly died from scarlet fever during an alert.

One night a fire bombshell fell on the roof of the house, Amama hurried out of the cellar to extinguish it with a sand bag (there were some all over). It was not the first nor the last...

One day at six o'clock in the morning, brown uniforms knocked at the door and ordered her to follow them immediately to the German police headquarters: "What is this letter ?" They asked her not too friendly. With her funny way of speaking German , she replied: "How can I know if you do not show it to me !... "It is probably from my daughter who is married to a British consul!".

"Never receive one again" , she was told.

"How can I tell her ?"

She went home with shaking legs, but remember: her ancestor was the Chevalier Bayard Sans Peur et Sans Reproche!

After that she sometimes got news from Jacqueline through the diplomatic bag.

The last alert, in 1943, chased us definitively out of the cellar on the Westwall in the middle of the night, the whole row of houses was burning, we splashed water from the fountain on our heads, hurrying away, leaving a heap of ashes behind. The next step was Cologne burning all around her beautiful cathedral (still there), escaping bombs in the overcrowded train to Friedrichshafen, by Constance Lake, where I remember seeing the Zeppelin factory burning and the sky glowing.

Eventually the war came to an end, for us as for many other German refugees, in a small village not far from the Black Forest: Aasen. There we survived a famous battle between the last of the Nazi army and our liberators, the French army. France had been liberated ten months before, when the Americans, Canadians and British army landed on the Normandy coast.

At last Amama could return to Paris to her sister Valentine at 55 rue de Bellechasse. We followed later on, my mother got married again to a nice French soldier she met after the famous battle in Aasen, Robert Claudé. He became the father of my two little brothers Peter and Jacques.

I lived in Paris until my marriage with Robert Astrié, a handsome French military doctor. Robert met John for the first time on our wedding day, such a special occasion...

John was very fond of our three children: Gabriel, Philippe and Caroline. He was godfather to Caroline.

We have now 13 grandchildren, which John loved too, he pushed their prams and skied with them in the Alps, many times, with all of them..

My brother Peter was a close friend to John too; they shared a lot, talking about science, telescopes, astronomy. My brother Jacques was the god-son of aunt Jacqueline whom he liked very much. John was very much interested in Jacques's family of five children he got from his three different wives. Poor dear, he would say he liked them all the same.

Unfortunately the link is lost with uncle Julien de Puthod and his family. Some in their family emigrated to America and Canada, John kept in touch for a while.

Do you understand now why John was at home everywhere ? Especially in our home, in Peter's and in

Jacques's.. ?! He listened to all the stories and witnessed each important moment and we all loved him for that.

I am sorry to say that nobody is left on the German side. Only some friends of mine and John's and a family he liked much who now lives in Chile.

I have met many relatives on the British side:

I have known dear old blind Lady Ponsonby and also her children: Uncle Arthur, Aunt Felicity and Uncle Jack Kennaway and their children, Jonny, Richard and Marygold, Aunt Catherine Ritchie and her children : Justin and Pam, their son Alastair and their daughter Siân (same age as my daughter Caroline) Uncle Gilbert and Aunt Sylvia and their children. And on the Buxton side, Henrietta and her three children (Augustin, Sophie and Thomas); two of whom spent time with my family during summer months. And Rosy Swale who was famous for sailing or running around the world.

That was the time when Uncle Arthur and Aunt Jacqueline retired to Thame.

Arthur and Jacqueline spent the end of their life happily in Thame at 26 Upper High Street, a nice little town in the middle of England near Oxford. London was not far to drive to as was Marlborough to visit John at his school. They were always very welcoming to everybody.

The French relatives used to visit them in Thame. My brothers and I spent several holidays there, John was not always at home, only for Bank holidays or summer holidays. We wouldn't see him much at the time, a pity, much less than later when he was grown up.

Aunt Jacqueline used her Montessori gift to teach French to English children and English to us French children. I can tell you how fun we had in Thame.

When John started teaching and doing research at Jodrell Bank, he bought his home in Wilmslow. We visited him on several occasions over the last 50 years.

PARENTS

Arthur Gordon Ponsonby

Summer Fields

Summer Fields is and was a preparatory school situated just out side Oxford. It was established in 1864 and took day-boys and boarders between the ages of eight and fourteen. Its *alumi* include, from around Arthur's time, Henry Mosely the scientist, Gubby Allen the cricketer, Prince Rainier of Monaco and, a little later, Lord Lucan. Arthur was a year behind Harold MacMillan, the future Prime Minister.

The only record of Summer Fields originates from an index compiled by the Old Summerfieldian Society in 1959. It records that Arthur played for the cricket first eleven from 1905 to 1906 and the football XI from 1904 to 1906.

There is little information about his time at Marlborough and Trinity

On the outbreak of WW1 both sides promptly interned foreign nationals who were present in their countries. The racetrack at Ruhleben near Spandau, Berlin was swiftly converted into a prison camp with internees accommodated in the stables and haylofts. By October there were just 80 or so occupants, mainly civilian British fishermen who had been picked up from trawlers which were sunk in the early days of the war. However they were joined by civilians who were picked up in mass arrests starting on 6 November. It is likely that Arthur, who was taking a holiday in Germany when the war started, was part of this.

The population of the camp grew rather quickly to some 5500 souls. There was a police force composed entirely of internees. Football and cricket teams were formed from a few professional sportsmen among the inmates along with keen amateurs. The Germans initially tolerated all this, then banned it and then allowed it again. It gave some of the internees something to do. Arthur played cricket for the Varsities team against the Rest of Ruhleben in 1915 . In 1916 his eight innings brought him an average of 55 runs.

There was a Ruhleben Horticultural Society who grew veg to supplement the cruel, sparse diet in the camp. In this they had the support of the Royal Horticultural Society who provided advice and even seeds. The cultural life was quite rich, virtuosos gave concerts and several operas were staged. There were educational activities and a library. Arthur spent a lot of his time there learning languages that were to prove useful in his future career.

With his all-male harsh public school background and Trinity behind him he no doubt took to the rigours of internment rather well.

There were several attempts to get Arthur released but all to no avail. He had to wait until war's end in 1918.

He joined the civil service at the Foreign Office in the Consular Service on 23 February 1920 as a Vice Consul. His first recorded posting was as an (acting) Vice Consul in New York in 1925 for two years. He was made then a Vice Consul there for another two years until 1928. The next posting we know about wasto Paris from around 1933[10] to 1937.

We next find him living in Beira, Mozambique when he married Jacqueline in 1938.

Marina gives this account of how they met:

> His [John's] parents, Arthur Ponsonby, his father and Jacqueline Kirdorf, his mother, sister of my mother, met in Paris.
> At that time, Arthur was working at the British Embassy in Paris. He used to go to tea parties at 55 rue de Bellechasse in Paris. There lived Jacqueline's aunt, tante Valentine (my grand-mother's sister).

10 He went to Rio in October 1933. Charles, Winifred and Priscilla Ponsonby were on the same ship but were no obvious relations. We don't know how long he stayed or why he went there.

Every Monday Valentine would invite young people, friends of her nieces and of her own, from all over the world. These parties were known as "les lundis de Valentine".

John has a slightly different one:

> Marguerite de Leusse, Viscomte de Puthod (but Daisy to the family)[11] was given to elegant dinner parties at her Paris home and in 1934 she invited Arthur Gordon, then a vice-consul in Paris, to one of them. Also there was Daisy's niece, Jacqueline Kirdorf. The two of them seemed to hit it off and must have seen one another in Paris (according to John in a letter to Renate).

When Arthur was posted to Beira in Mozambique in round about 1935 or 1936 they must have kept up the relationship by correspondence because in 1937 he took a hesitating step and proposed marriage .

John indicates that there was a degree of hesitation:

> My father proposed marriage by letter which he sent back to England by diplomatic bag. He thought better of it and tried to get the letter intercepted at the Foreign Office in London. But they failed! So mother came out to Beira[12].

While a British Consul can conduct marriages he cannot perform his own. So the two of them took a train to Bulawayo (now in Zimbabwe but then Southern Rhodesia) to be married by special licence on 5 February 1938· By then Arthur had a new posting as Chargé d'affaires in Monrovia, Liberia on the west coast of Africa. So, after the wedding, they set off overland and by boat on the Congo on a journey and honeymoon to set up a new home together on the other side of the continent.

11 John thought her a "dreadful old snob"

12 She boarded ship in Genoa in December 1937.

Within just a few months, Jacqueline became pregnant and it was decided that she should return to Europe to have the baby. So, later in the year she travelled to England (via the Cote d'Azur) where John was born, in South Kensington, on 25 March 1939.

Arthur stayed on in Monrovia[13] until 1940 when he was recalled to London. There were suggestions that, since Jacqueline was German, it was inappropriate that he should serve in wartime in a normal consular position.

According to the letter from John to Renate:

> It was a bad time of course just before the war and mother had been a German national till she married (I have her German passport). But the German consul in Monrovia had recently married an English girl. So after the war had started and they had to be very distant with each other on public occasions, in private they used to have tennis parties together!

He worked in a department dealing with Allied Prisoners of War until 1945.

Illustration 2: Newborn John and his parents

13 Although he made a visit to London because there is a record of him sailing back to Monrovia on 30 August 1939 – just two days before World War 2 broke out.

Illustration 3: John and his Father

Harry Pemberton Ponsonby

It was during this period of stable home life that his mother asked him to sort out the affairs of the intriguing Harry Pemberton Ponsonby, who had died.

Harry Pemberton was born in India on 15 October 1876 to Charles John Ponsonby and Isabella where Charles worked in the Forestry Department. By 1891 Harry was living with his cousins George Montague and William Randall in lodgings of some kind as Scholars in Sedburgh. By 1904 the family was living in London. Harry somehow ended up in South Africa as a Burgher in the Waterberg Commando in the Boer War. He allegedly changed sides and enlisted as a Private in the 6th Inniskilling Dragoon Guards.

After the war finished he went back to India until he returned to the UK on the outbreak of the First World War and enlisted in 1915 as a Lieutenant in the Royal Field Artillery(Territorial Branch) in the 4th East Anglian Brigade. In 1917 he married Elsie Myell and had three sons: Douglas Henry Talbot, John Gordon and Charles Malcolm. He retired from the army in 1921 with the rank of Captain.

He lived in various places including Harrogate (where he was on the Freemason's Register) and Frinton-on-Sea, ending up in the Watford area.

In a Will dated 27 August 1930 he left his entire estate to be divided equally between Elsie and his children. The estate included, intriguingly, "my rights in patents plays and other writings".

He was the first one to apply for a place in a charity development at Macdonell Gardens in Watford built for the wounded in the World War which opened in in October 1932. The criteria for acceptance were that candidate should be partially disabled, married with at least one child , and of the "right type". They should be able to support themselves financially. Quite how Harry did this is not certain but in the 1939 Register he gave his occupation as "Author".

Harry not only applied for (and got) the first bungalow in the "Watford Colony", as it was known locally, but he took on the unpaid job of first Agent possibly because he had offered to dig a well for the community. The community comprised eventually 12 bungalows – but Harry never dug the well

He was known to all as Major Ponsonby. Whether his rank was inflated by some official means or whether by Harry's prompting and demeanour we can't be sure.

He dropped the Agent job in 1934 and Elsie and the family seemed to be happy there until 1941 when Harry became ill and was forced to move into a nursing home sometime in May or early June.

He enters this narrative because he was Arthur's first cousin and Arthur became involved – at his mother's instigation. By the end of June Harry was much weaker and had difficulty breathing. On 26 June he sent a postcard to Arthur in a pencilled scrawl:

> Will you please be my Executor
> Moriturus te saluto

Harry

Harry died on 8 July 1941 just a couple of weeks after his appeal to Arthur.

The two immediate concerns, after giving Harry a decent funeral, were that Elsie might be ejected from her home and that her finances were in a poor state. Arthur managed to get assurances from the charity that nothing precipitous would be done but that at sometime the criteria for residence would have to be applied. The couple were really very poor but there was a life assurance policy which paid out a few hundred pounds and the situation was (rather sadly) eased when it emerged that Elsie's mother, who had recently passed away , had left her about £690 in her will.

Elsie moved away and died aged 84 in 1980 in the West Country.

After the War

In April 1945 Arthur, Jacqueline and John set sail from the Clyde to Port Said in a convoy (the war was still going on and the U-boat menace was not over) to a new posting in Alexandria. As they entered the Mediterranean the Captain announced on May 7th that Germany had surrendered. The convoy broke up and, after a night at anchor in Gibraltar (John remembered the lights) they pressed on, in lighter spirits, to Port Said. They took a train from there to Cairo and then to Alexandria and there they stayed for about ten months. John went to a local school and visited some of the battlefields in North Africa. He was only six years old; some of the images of devastation were to stay with him for many years afterwards.

In March 1946 Arthur was posted as Consul to Luanda in Angola and they were off again: Suez and then by ship to Mombasa, Durban and Lourenco Marques (now Maputo). Here John had his 7th birthday in a hotel (his present was a wind-up toy train). They then set sail, this time on a Dutch ship, for Cape Town (they visited the Simonstown naval base) , Lobito in Angola and then finally Luanda, the capital. It was a long voyage and John recorded that he "learnt to play chess."

Arthur was in Angola for just about a year; Jacqueline and John spent six months of that in South Africa as a result of Jacqueline's illness (see below) and then they returned to the UK.

Arthur's final posting was as Consul General in Rio de Janeiro from 1947 to 1951. He then retired in that November and he and Jacqueline soon took a house on Upper High Street in Thame, a town 13 miles east of Oxford

He died in Victoria Cottage Hospital, Thame on 18 April 1978 from pneumonia following a stroke.

Jacqueline Ponsonby

Jacqueline Ursule Valentine Bertha Margarete[14] Kierdorf was born in Krefeld in Germany on the 21st of April 1904. Her father Carl Julius Kierdorf[15] was a paper merchant and her mother, Ursule, was from the de Puthod family, one with rather faded aristocratic credentials.

They had two other children, both girls, Elisabeth and Claude.

Arthur's strange proposal of marriage, so seemingly romantic and so nearly withdrawn, seems to have been a sign of future strains in the marriage,

As John recalled many years later:

> I think they themselves had a funny relationship. I never saw them showing any demonstrably loving contact, hugging or anything like that. I remember my French cousins remarking that they couldn't understand how my parents ever managed to produce me!

John wrote in 1996:

> As for my mother I guess I loved her but I am also sure that in a way I didn't particularly like her. She was rather jealous

14 There were numerous other forenames, too many to mention. She preferred to be known as Jacqueline Margaret

15 "Kierdorf" at some point became "Kirdorf", perhaps because it sounded less Germanic

of the fact that I so enjoyed my father's sister Katherine. Mother had many attitudes to life which I did not like. She was over formal with people and was terribly concerned with appearances. You will have noticed that I don't care a fig for my appearance except for occasionally getting my hair cut! Katherine was like that. She could make friends with anybody. In her mid seventies she allowed herself to be picked up in the street by a Chinese student who came up to her and said "You have an interesting face, may I photograph you?" He was studying photography and his pictures were marvellous. Mother would never even have spoken to him. I don't know if she showed her love for me. I'm not sure in what way you mean. Certainly she was proud of me to an exaggerated extent that was sometimes embarrassing and she always wanted to give me things. All too often they were things I didn't want but which she thought I ought to have. I felt she tried to force her values on me. So I don't know. I've only had one mother and I don't know how she compared to others.

Breakdown

The first mention that Jacqueline was in a clinic in South Africa was a letter of 18 September 1946 from Marjorie Byron to Arthur. It seems that she had had a nervous breakdown while in Angola and the Angolan government provided a plane to get her to South Africa. The clinic was the Florence Nightingale Nursing Home in Johannesburg which specialised in "Nervous and Mental Diseases". By late December Dr Geerling told Arthur that the treatment had been completed and that Jacqueline had responded well and there was a big improvement and she seemed stabilised. He was thinking of allowing her to visit a cinema with one of the nurses as part her rehabilitation. She felt better.

We don't know when she left the clinic but Arthur left Angola, presumably for South Africa, in mid-February and it would seem likely that she was discharged around this time. John left his school (Park Town) in early March and they all came home shortly after that. John, as we shall see when we come to his education, was not on the plane that evacuated Jacqueline. He had another adventure altogether

Her interests included collecting lace and her considerable collection was donated to the Lace Guild's museum in Stourbridge near Birmingham. It is still known as the Ponsonby Collection.

She died in Thame on 15 August 1989.

3: JOHN'S EARLY YEARS

John was privately educated, generally as a boarder, at a series of nursery and preparatory schools before going to Marlborough College. Until Marlborough, it was a rather disjointed affair.

Wansfell[16]

John was sent to Wansfell School in St Albans in 1944 and stayed there for about a year as a day pupil.

He was obviously a keen child : the words good and attentive crop up in most subjects in his term report. In Scripture he was very good and attentive, his Writing was "very neat". In "British History" he enjoyed the stories and "narrates quite well" while in Arithmetic he had "a good idea of the value of numbers". In "Drill" he was very good and joined in well. His "Work" was very intelligent.

His "Conduct" was "Much improved during the term".

In his second term he managed to impress with his quickness and made progress in Arithmetic. He was now recognised as "Very good " at Handicrafts. His "Conduct" he was "still too noisy in the cloakroom but otherwise much improved."

In his third term he was generally interested and shone at Handicrafts with a "Very good indeed". It comes with a hint of inconsistency: "does not always try" in "Reading". The comment on his "Work" is "good, but it varies very much." "Conduct" was censured as being "too noisy and rough this term".

Something of note that is not highlighted was his attendance

---

16 Wansfell was a PNEU (Parents' National Education Union) school – probably located at 76 Clarence Road St Albans. PNEU was an organisation promoting a liberal Christian education.

record. Over the three terms he was present on only 57% of occasions when he should have been.

Alexandria and Homecraft

John left Wansfell because his father had been posted to Alexandria. He was on on a ship in a convoy from Glasgow bound for Cairo with his father and mother. When they docked in Cairo they made swiftly to Alexandria and that September John started at his new school.

He was at Homecraft Kindergarten at 7 Rue Pharaous for just one term.

The Handicrafts teacher said "he works enthusiastically". Singing was "very good indeed" but Dancing just "Fair".

General Progress was "fairly good on the whole but spasmodic". But there were encouraging Remarks from the headmistress Miss Gribble[17] : "A very intelligent and promising pupil whom we are very sorry to lose."

He was again absent rather a lot: 35 times in the term.

Park Town

John was at Park Town School, Mountain View, Johannesburg (PTS) for the final term of 1946 until the first term of 1947 – so just six months or so.

It seems to have come about because Jacqueline had been admitted to the Florence Nightingale Nursing Home in Johannesburg after a nervous breakdown.

> I think I must have been 8 when mother was suddenly taken ill and was flown to hospital in Johannesburg. I was dumped (it took ten days by ship and train to get there) in a boarding school, Park Town School, in the suburbs of Johannesburg with instructions to write a letter once a week to my grandmother in London. It was not a happy way to leave

17 Miss Gribble was related to Leonard Gribble, an astonishingly prolific and successful writer of crime and western novels.. A founder of the Crime Writers' Association, his work *The Arsenal Stadium Mystery* was made into a film

home.

In fact it must have been some adventure. Luanda to Lobito by ship and then train on the narrow-gauge wood-burning Benguela railway to the Belgian Congo (now the DRC) and, via Southern Rhodesia (now Zimbabwe), to Johannesburg and Park Town School (PTS). The train journey alone was something like 2500 miles.

He must surely have been accompanied but there is no mention of it.

Illustration 4: The family in South Africa with John in PTS uniform

The school was recommended to Arthur by a friend. Correspondence between the headmaster of PTS and Arthur revealed that the school would be delighted to have John but they would rather have nothing to do with his mother. She would not be permitted to visit the school, the school would not be available to John in the holidays and they could not help with finding accommodation for Jacqueline when she came out of the Clinic.

The general tone of the report for the first term was satisfactory, although good in arithmetic. In Games he was "Vague". His Conduct was "Fair: has found it difficult to mix with the other boys."

In the second term his Reading, Spelling and Writing were, variously, slow, very slow, weak, untidy and inattentive. In Scripture he acquired a "Very good". The assessment of his Games performance hardened up to "Sometimes did quite well". He seemed to be settling down according to the head.

We can't be sure but it seems likely that arrangements were made for John to go to Grahamstown for the Christmas holiday. There is certainly a letter from an L M Beaufort at The Deanery there on 17 December describing how John is settling down and another from Ruth Beaufort at Bushmans River Mouth, on the coast, on 6 January 1947 saying that John is enjoying the break, swimming and learning to fish.

In a letter to Arthur in April the head said of John:

> He was, through force of circumstances, a difficult chap to deal with. But he was getting more used to living with other boys, and a good deal happier here than he would care to admit.

John left the school in March 1947 because his parents were returning to the UK.

Kestrels

Kestrels was a school in East Anstey, North Devon. Requisitioned for a Surrey Prep school during the war, the owner, Richard Stapledon, after it ended decided to set up a school himself and become its Head. The school survived until the mid-60s.

John was there from the summer Term of 1947 and left after the Spring Term of 1949 when he moved to Ashdown House. He was fairly consistently in the middle of the form (which seems to have varied from six to eleven pupils); after his first term though he came second when his Mathematics drew a remark from his teacher of "Brigh(*sic*) and intelligent".

In his second Term he played rugger for his House and his Riding had improved. However the Head's remarks were "For his own good though, John, must give quite serious thought to the control of his 'feelings' and curb them."

By the next Term John is complimented at the way he had tried to curb his feelings but "He is still rather inclined to consider 'self' before House..."

By his final Term he was placed first in his form for Mathematics but 9th (out of 10) for Divinity – which, rather puzzlingly, merited a "Good" from his teacher.

Stapledon, in his Report, warned that "If only he could and would realise that a thorough grasp of essentials and fundamentals is a necessary stepping-stone to success, then his development would be more rapid. He has ability." He said that he regretted John's departure "more than I can say at this most important and impressionable stage of his life.

Ashdown House

John seems to have had some time off at the beginning of the Winter 1949 Term, possibly as a result of the trip to Rio and he missed the exams. The Head's report suggests that all was not well. John was:

> ...not strong constitutionally. This weakness leads to fatigue and fatigue to a certain aggressive cantankerousness. Then sparks fly! He is inclined to be intolerant and self -assertive...

Reassuringly, in the Summer Term there were "few outbursts of temper". Spelling was a weak point still but in Mathematics he came top in the exam with a score of 100%. He followed it up in Lent Term by coming top of the class of 13 boys. The Head thought he had a decided flair for the subject.

He came first in the Winter Term too.

In conduct he was much improved with only one tantrum – where he had been "as much irritating as irritated" in the past. He was getting on with his contemporaries much better, "taking criticism less resentfully."

He was placed 12th out of 12 in Latin, French and English. He still excelled in Drawing and Handwork but was “not very willing to take advice.”

John arrived almost exactly when W G Williamson took over. Williamson reigned for 25 years. A classic scholar, he was known as Billy by the boys and ruled with a rod of iron. Beatings for the smallest offences were common both by the staff and the older boys. Williamson himself seemed to relish administering them. As well as corporal punishment (which was legal at the time) there were allegations of sexual and mental abuse.

The alumni included Viscount Linley (Princess Margaret’s son), Boris Johnson, Andrew Mitchell MP, Damian Lewis, the actor, and a future Metropolitan Police Commissioner.

The school closed in 2020.

Illustration 5: John on a pony at Ashdown House

Marlborough

We have his school reports from 1953 until he left in 1957. I have selected a few comments made in each year. Those from his House Master are labelled HM, those from the Master (the Headmaster) H. Otherwise they are from another teacher.

1953 "He creates difficulties for himself with his quick temper and prickly ways... He must make efforts to become more sociable... he stamped along on his own about 20 yards ahead" (HM).

"Temperamental and prickly on the outside but sound underneath"(HM). His spelling was "<u>very</u> bad".

He was moved from Upcot House to Cotton at the beginning of the 1954 school year.

1954" I'm rather disturbed about the standard of his work and lack of progress. In his relationships with others he is his own worst enemy because of his fiery temper" (HM).

"He can be most charming and cooperative person , then he suddenly flies off the handle",

Things seemed to be improving. He was top of his set in Physics

"There is a noticeable all-round improvement" (HM).

"By no means discouraging" (M).

1955 " Seems to be generally working harder."(HM)

1956 "He is rather slow to see some simple things but at the same time is quick to spot one or two points that pass over the heads of others". (Referring to maths -- HM).

"His reports are always very much the same" (M).

"A determined worker who uses his abilities to the full. He has made good progress throughout the year"(M).

"A thoroughly sound and intelligent boy. He is developing well."(HM).

In Latin he was still hopeless: " useless".." going backwards...he has given up the unequal struggle"(HM)

1957
"He has been an active and cooperative member of the form. He has ability and has done much good work."(HM)

"I was particularly impressed by his railway exhibit on Prize Day."(HM)

In English: "satisfactory but in no way promising".(HM)

The Master's final comment, as he passed John on to the outside world was: "His ability lies more in practical than in theoretical world and I have no doubt that he will be successful". It was either incisive or very wrong.

One notable thing is that the comments about his temper and unsociability ceased after the first year. He may have found some like-minded friends. He certainly joined the Radio Society and started his life-long obsession with scribbling electronic circuits on odd scraps of paper He may have benefited from the character and team building experience of becoming involved in Fox's Folly – the bridge built by a group of boys. The change of house after his first year may have helped too.

Illustration 6: Fox's Folly

COTTON HOUSE TEA

Illustration 7: Tea in Cotton House

Encouraged by Ian Browne, a teacher who had some knowledge of the Jodrell Bank Radio Telescope, John developed an interest in astronomy. He passed six subjects in GCE 'O' Level exams in the summer of 1955, followed by Chemistry and Latin the next year. He took his 'A' Levels in 1957, achieving a Distinction in Applied Mathematics

Illustration 8: John in 1955

Imperial College

Almost nothing survives from his years at Imperial.
He graduated with a 2(1) in electrical engineering in 1960.

Escot

Escot played such an important role in his early years that it seem appropriate to regard it as part of his education. The below is taken from an email John wrote to Irma Kennaway:

> I haven't got much to say about Escot pre 1947. I know that prewar and in 1939-40 or thereabout every child at Escot had its own nanny. Before the war I gather the house had a staff of butlers and chamber maids and cooks, the lot, but I have no memory of those times. That all came to an end at the beginning of the war. My mother left me at Escot with a nanny at about Christmas 1939 during the "phoney war" and before the fall of France. She returned nine months later in September 1940. I have a picture in my mind of meeting her at Sidmouth Junction wearing a white coat or outfit. She was astonished, having left behind a baby, to find a small boy walking down the platform towards her. But I'm not sure this is a real memory or whether it is an image created by what I was told. Somewhere I have photos of the Escot nannies. I'm not sure I can find them but they were certainly a grim faced looking lot. I have for long wondered what mischief they got up to. Did they abuse us in some way? (I once tried to sound out Richard on this theory but he didn't respond.) I am told that they used to wheel us in our prams to the level crossing on the road to Ottery in time to see the 3 o'clock cho-cho go past every afternoon. The old level crossing gates are still there in the hedges. At some point Escot House was taken over by the Dr Bernardo's*(sic)* children's charity and we all moved into Fairmile House. I guess the nannies left at that point and my mother (Aunt Jackie) and Felicity looked after

the four children. In order of age Johnny[John Lawrence], Richard[Richard Noel], JB, Marigold [Mary Joyce].

I think Felicity and old Sir John (Jack) were an unlikely couple with no discernible interests in common. How they met I don't know but I do know that the Kennaways and the Buxtons had been associated in some way for a long time.

I have better recollections of Escot after I returned from Africa in 1947. Then I went to a prepschool (my second) in north Devon on the edge of Exmoor (Kestrels at East Anstey) and spent all my holidays at Escot whilst my parents were in Rio. By then I had a camera and I have my photos. Most of the house was unused and Marigold and I sometimes made explorations into the unused parts and clambered about outside on the roof. I recall that at some time when my parents were there and there were then nine people living in the house a cook was employed. None lasted long. I remember that after one of them left we found lots of baking trays full of rice pudding that had got burnt in the oven. She had been too ashamed to admit it and we found them hidden all over the place!

It was at that period that the Noels lived in the flat upstairs. I remember Gerard and his sister Celestria and their mother. She taught me how to make porridge, a skill I exercise every morning. I don't recall whether Gerard's brother Methuen was there. He died very young.

4: JODRELL BANK

In the October 1960 John registered for a Ph.D. degree in the Department of Physics and Radio Astronomy at Manchester University and joined Professor Bernard Lovell's group at the Jodrell Bank Observatory.

He was essentially an electronic engineer with an overwhelming interest in processing the signals picked up from the rest of the universe by radio telescopes to get the maximum information from them. He was not as passionate about understanding the processes that generated the signals – the astrophysics.

The first thing he was asked to do at Jodrell was to join a team using radar bounced off of the planet Venus – not quite in the mainstream of what was going on at the time. He was to spend just about a decade doing this. At first to measure the Astronomical Unit (the distance of the Earth from the Sun) and later to measure the rotational speed of the planet.

Venus Radar

Kepler's Laws give a remarkably accurate account of the motion of the planets of the solar system provided that we have just one actual measurement. The distance between the Sun and the Earth (the Astronomical Unit – or AU) is the one usually taken but any other planet would do. Historically the main way of doing this is to measure the transit of Venus. A transit occurs (rather infrequently[18])

18 The pattern of the transits repeats every 243 years and comprises two in June separated by eight years , a 105 year gap and then then, again, two in December eight year apart followed by a longer gap of 121 years. We are at the time of writing between two December transits: one on the 10-

when the planet, visible as a black spot, crosses the disc of the sun. Timing how long it takes to go all the way across at various observation points on the Earth's surface gives, with some geometric analysis, the AU.

Radar allows us to measure the distance to our nearest planetary neighbour when it is at its closest in its orbit (at inferior conjunction) and for just a few days.[19] Using this information similarly allows us to derive a value for the AU.

The Jodrell 1961 System

The 1961 system was a pulsed one that timed the pulse from its transmission to reception back at Earth. The results gave a value for the AU which was significantly lower than the more recent ones. Why this was so was never really explained. However, when it came to designing a system for use in the next close approach in 1962 the system chosen was radically different.

The 1962 System

The system was constructed between the inferior conjunction (closest approach) of April 1961 and that of 13 November 1962 and it was to be a continuous wave (CW) radar to measure the closing velocity of Earth and the planet using the Doppler effect rather than a pulsed system to measure the distance between them. Since the same aerial (the large Jodrell dish) was to be used for transmitting and receiving, the transmit cycle lasted for several minutes, then there was a pause and then the receiver was switched on to wait for the returning signal. The cycle was repeated every few minutes, allowing time for the transmitted wave to get back.

Also, since the AU was know quite accurately by this time, this value could be taken and the velocity of approach of the two planets could be predicted using Kepler's Laws. They decided to use the value

---

11 December 2117 and another on 8 December 2025.

19 Mercury and Mars have been detected by Earth-based radar as have various passing objects, including the Moon.

149 589 500 km determined by JPL[20].

To get the closing speed using the Doppler effect account also had to be taken of the effect the Earth's rotation – or rather the telescope's velocity towards Venus as the world went round.

These calculations (and others that would ensure that the telescope pointed at Venus throughout the observations) were done on the University's *Mercury* computer and stored on 8-hole punched paper tape for every five minutes of the observations.

When the transmitted signal was received back from Venus it was progressively stepped down in frequency to 1MHz and amplified. It was then combined in the Ephemeris Doppler Machine (EDM), driven by the paper tape, to remove most of the effects of the predicted velocity of approach and the Earth's rotation. It was arranged that if the predicted velocity was right the signal would be just 30Hz. More frequency reduction followed to produce an output centred on this value.

This was then passed through a digital frequency filter which was set to sort it into 11 channels, 1 Hz apart, from 25Hz to 35Hz. (For technical reasons things were adjusted so that, if the predictions of the EDM were correct, the returned signal would be centred on channel 8 rather than channel 6.)

The results were accumulated over several hours on mechanical counters

It was clear from these straight away that the peak response lay in channel 8 and that the value assumed for the AU was not far out.

To get the most out of their rather sparse data they then set about fitting a Gaussian distribution to the returns in channels 7,8 and 9 it to get a more refined value of the peak frequency and the half-width.

The peak fell consistently at a slightly lower value than expected but only by a fraction of a Hertz It meant that the AU was probably about 2000 km smaller than the JPL value used in the ephemeris calculation. That put it towards the lower end of the radar determination from 1961 – and about 60 000 km greater than the then currently accepted value from other sources.

The EDM was John's major contribution to the project. It was, as

20 The Jet Propulsion Laboratory

far as we know, his own idea, and it cleverly used the then-current technology to make the complex ephemeris calculations performed on the computer to keep the giant telescope pointing in the right direction and to correct the results obtained.

None of the other teams had anything quite like the EDM. They clearly managed to get results, but not quite as elegantly as John did.

Illustration 9: The Ephemeris Doppler Machine

They had hoped to draw some conclusions about the rotation rate of Venus. This might have been seen as a broadening of the spectrum of the returning signal. Half of the surface of the planet would be moving away and half of it moving towards the Earth if it was rotating. If it was rotating fast enough, it might show up as a Doppler shift.

Analysis of the results showed that rotation period of less than 255 days was likely either direct or retrograde. So in fact it did not add much to what little was known already.

The 1964 System

The 1964 system was very similar to the 1962 one except that a better RF amplifier was used. They managed to crack the channel 8 problem and the peak now appeared right in the middle at channel 6. The results looked a bit crisper and there were more of them as the data collection time was from 5 June 1964 to 28 July.. As the closest approach was on 19 June, they managed to watch the planet approaching and retreating.

The value obtained for the AU was substantially the same as in 1962. The measurement of the rotation confirmed that it was retrograde but analysis suggested that the period was 100 days rather than the 243 days which was by then the accepted value. More analysis with some challengeable assumptions about the noise spectrum gave 245+20 to -25 days. This encompassed the accepted value but was still a rather disappointing outcome. Plans to record a great deal of data on paper tape for later analysis by the Atlas computer were successful but alas the computer time was not available to do it.

A further disappointment was that the spectra showed no sign of any persistent features that might indicate rotation.

Russian Collaboration

The idea for collaboration with the USSR came from Lovell and Academician Kotelnikov of the Institute of Electronics in Moscow. It was that a 768 MHz transmitter (Pluton) at the deep space tracking station at Yevpatoria, Crimea, considerably more powerful (10-80kW) than that available at Jodrell, would be used with the signal detected at Jodrell using the 250 ft Mark I telescope and an enhanced version of the 1964 system. It was a bi-static arrangement (it used separate telescopes far apart) and so the signals could be transmitted and received continuously. The collaboration ran from December 1965 to March 1966.

The December observations failed to find the returned signal mainly because the ephemeris calculations were wrong. Even after these were corrected the signal could not be found. It was only after a laboured search of nearby frequencies that it was found on some 124 Hz (a large error in the circumstances) below the expected frequency. The planet was so close to inferior conjunction that the signal was quite strong. They then could continue in the normal way – except that on each occasion they had to search for the signal. They thus made contact for about two hours on each of eight days in the next two months. The final contact was on 11 March when it was decided that the signal was so weak that it was not worthwhile to continue

The problems were never fully followed up but John was of the opinion that the error was in the transmitted frequency and was astonished that the only deep space tracking station in the USSR had far worse frequency control than was possible using commercial equipment available in the West.

The width of the returned spectrum was around 11 Hz. This was consistent with the accepted value of Venus's rotation rate at the time of 243 days within an error of about +/- 5 days. It had been hoped that something in the spectrum might be found that indicated a feature on the planet's surface. Some of the spectra obtained were slightly asymmetric and, on one or two, there seemed possibly to be anomalies associated with features already found by other workers.

John later suggested that the results of the collaboration were not really worth all the effort and that it was all done for political rather than scientific reasons.

PhD

John was awarded his PhD in 1969 for the radar work

In his summing up in his thesis John expressed some disappointment:

> The scientific results from this work have been disappointingly meagre. Planetary radar has proved to be a field in which new results are only obtained by the groups which have the most sensitive systems and the data

processing capacity to make the best use of the data acquired. in both respects the group at Jodrell Bank has never been in a leading position. Indeed it has only been extreme exertions that have made it possible for a group consisting of seldom more than two or three people to obtain echoes from Venus at all.

As a consequence the results obtained have invariably been confirmatory of the work of others and were generally speaking one inferior conjunction behind.

At the beginning of the decade it was clearly not appreciated how expensive the most powerful transmitters were to become and no attempt has been made to acquire one which will do justice to the 250 ft Mark 1 radio telescope. Instead the great advantage of the large instrument which has been which was considerably bigger than anything available to other groups was used to make up for this very modest(5 KW) transmitter

If a true state-of-the-art transmitter were acquired it would cost an appreciable fraction of the cost of a telescope and clearly to justify investment on that scale it would have to be used much more extensively than would be compatible with the predominantly passive radio astronomical programme at Jodrell Bank. Passive radio astronomy may appropriately be done as a secondary line of research at a primary radar installation but experience has shown that the two activities do not combine well the other way. Appreciating this, the research reported in this thesis is not, at least for the present, being pursued further. Though it suited the predilections and interests of the author to have to design the entire radar system, the modest level of technical assistant available made it impossible to attempt more elaborate observations such as using phase modulations for obtaining range information with the basically CW system. The limited computing facilities available in the University and the lack of the time of on-line computers at Jodrell Bank in effect prevented a thorough analysis of the data that was acquired, and this took away much of the value of the observation. The magnetic and paper tapes containing 1964 and 1966 data have been preserved however and should interest return they could be analysed in the future when data processing capacity is more readily available.

Apart from certain engineering contributions of which the Ephemeris Doppler Machine and the Resistor machine are perhaps of most lasting value, the chief results of this work are the measurement of the A.U. as

149 598 500+/- 900 km

obtained in 1962 and the confirmation of the retrograde rotation period of Venus. From the measurements made by the author at Jodrell Bank this is found to be

- 243 +/- 5 days

Illustration 10: In PhD robes 1969

APERTURE SYNTHESIS

In optical astronomy obtaining an image is relatively simple.[21] Get a good telescope with a large enough aperture, fix a photographic plate at its focus and expose it for a long enough time, develop the plate and you have your image.

In radio astronomy, for best resolution, you need an array of telescopes spread out over a large area (perhaps even on different continents) all pointing at the same area of the sky. You then have to collect the very faint radio signals from each of them at some central point, preserving their phase relations. You then need a powerful computer to combine them, using some trusted algorithm, to produce an image.

John was at some point involved in most of these steps and I have picked out some examples of where he succeeded rather brilliantly and one or two where he probably failed.

Some of these are extremely technical. I have tried to make them comprehensible to the general reader without an electronics background but I am sure that, in some cases, I have failed. Where I have failed I hope that I have done enough to persuade you that John had, at least, a remarkable mind.

I have described one project he was involved in (the autocorrelator) in some detail in an Appendix.

So John said good bye to radar and joined the mainstream work of Jodrell. To understand what came next we need to understand the principles of aperture synthesis and interferometry which had such a dramatic effect on radio-astronomy.

An important parameter of telescopes, optical or radio, is their resolution – their ability to separate closely-spaced objects in their field of view. There can be several reasons for this being limited (such as turbulence in the atmosphere or poor quality optics) but there is

---

21 My apologies to optical astronomers for this travesty.

one fundamental limit on what can be achieved with a particular telescope and it's because light and radio signals are both electromagnetic waves. As a result of this the angular resolution of a telescope is limited to 1.22λ/D where λ is the observation wavelength and D the diameter of the collecting lens or mirror. The units of the resolution are then radians.

In arcseconds[22] that is about 2.5×10^5 λ/D.

For visible light with a wavelength of 500 nm this means that a telescope with one metre diameter objective would be able to resolve things with a separation of around 0.1 arcseconds. That's about the width of a human hair at 100m or, in astronomical terms, a few times the angular diameter of red giant stars.

In contrast, radio waves used in astronomy have wavelengths from around 1 mm to around 1m. The largest single-dish radio telescopes today are around 100m in diameter so their resolution at 1 cm wavelength is about 25 arcsec, much worse than their optical counterparts[23].

Interferometers take advantage of the wave nature of electromagnetic radiation. Young's double slit experiment in 1801 showed that if a beam of light passed through two narrow slits close together, it recombines on the other side to give a series of bands or fringes. The explanation was that light is a wave and the two waves passing through the slits cancel one another out in some places and reinforce one another in others, they interfere with one another. Hence the bands.

In astronomy the technique was used by A A Michelson in his stellar interferometer, a version of the double slit experiment which involved the 100 inch Hooker optical telescope. He managed in the 1920s to measure the diameters of several red-giant stars which

---

22 Angles are measured in degrees, minutes (1/60 th of a degree) and seconds (1/60 of a minute). The measurements are often expressed in arcseconds or milliarcseconds. Several abbreviation are used : arcsec, milliarcsecs and mas.

23 The only one bigger than 100m is the FAST telescope in China which has a 500m dish.

could not be resolved by the telescope in normal use.

The first use of interferometry in radio astronomy was by an Australian group. They studied the sun's radio emissions by combining the direct rays from the sun with those reflected from the sea using antennae located on the top of Dover Cliffs near Sydney. They showed that there was a marked non-uniform distribution of emission intensity across the Sun's surface, contrary to what was generally believed at the time.

The team's inspiration came partly from wartime radar work that suggested that the altitude of planes, flying over the sea, could be estimated from the interference patterns.

The significance of all this was that if two radio-telescopes could be pointed at the same distant object their resolving power becomes determined by their distance apart (D in the formula above) not their individual apertures provided that their signals are combined in the right way. Then they would have the resolving power equal to that of a single dish with aperture D – although not its sensitivity.

The first instrument which exploited the principle was an arrangement, built at Cambridge by John Blythe in 1954. It had about 2° resolving power at a wavelength of 7.9m and produced some of the earliest detailed maps of galactic emissions. It was followed by other arrays with improved resolution of 45 arcmins at 7.9m and 25 arcmins at 1.7m built by Martin Ryle and his team.

In 1960 Ryle and Hewish published a paper[24] "The Synthesis of Large Radio Telescopes" which is often seen as the single greatest advance in the early days of aperture synthesis.

The principles are fairly simple – although the details are complex. We will use the example of a two element interferometer – two identical radio telescopes (A and B) pointing at precisely the same patch of sky (centred on the radio source So) but a distance L apart. To do this, they are both inclined at an angle φ to the horizon. The arrangement is summarised in the Figure overleaf.

---

24 "The Synthesis of Large Radio Telescopes", Monthly Notes of the Royal Astronomical Society vol 120 p230

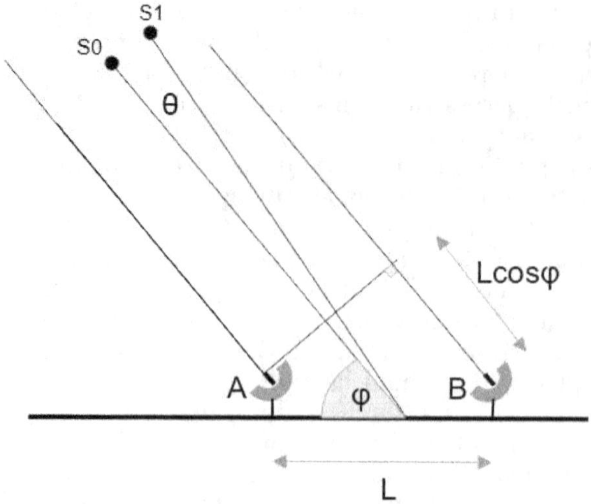

The radio waves have to travel further to reach B than A by Lcos(φ) or Lcos(φ)/λ wavelengths, where λ is the wavelength of the signals. They will therefore arrive with a delay of Lcos(φ)/c where c is the speed of light. This is the geometric delay. If we electronically delay the signal from A by this much and take account of any other delays that might arise in the system, the two signals from So will be precisely in phase. However, a bit more geometry shows that a nearby source S1 located at an angular distance θ (we're talking arcseconds here) from so will arrive at B with an even greater delay compared to its arrival at A. The signal at B will be phase-shifted compared to that at A by Lθcos(φ) wavelengths where θ is measured in radians. So the phase shift gives us a direct measure of the angular separation.

As a concrete example, a θ of one arcsecond is about 5E-6 radians so a baseline of 1 km and a 2m wavelength would give a phase shift of $5x10^{-6}*1000*0.5/2$, so about 0.001 wavelengths when φ is 60°. At 1cm wavelength this would be 0.25 wavelengths.

What about the rotation of the earth you may ask. Well, provided you keep the telescopes pointing the telescopes at So and you adjust the delay to match the geometric delay, it all works. There may be adjustments to be made to the calculations of the θ to phase shift but that is relatively straight-forward.

This has been described for a string of sources, the one dimensional case, but the extension to a two-dimensional array of sources is in principle just the same – only more complicated.

The next real problem is measuring the small phase shift and this is done with a device called a correlator. At its simplest a correlator multiplies the two voltage signals from the telescopes (after one of them has been delayed) and takes an average of this over many cycles. This average is directly related to the phase-shift and hence to the location of the sources.

To get information about the 2-D location of sources you have to have not just telescopes along an E-W axis but also some with a N-S alignment. Many arrangements have been tried. Some of them have telescopes on railcars so that they can be arranged for optimum performance because it turned out that wide spacings reveal more detail of of smaller heavenly structure but shorter spacing reveal more of larger ones. Each pair of telescopes form an independent baseline and have to be fed into a correlator.

The largest purpose built array in the Very Large Array in New Mexico which has 27 telescopes on a Y-shaped rail track. It has 351 independent baselines, receives at wavelengths between about 0.6 cm and 410 cm and is capable of a resolution of around 0.1 arcseconds

Over the next two decades many arrays were built with longer baselines (hence greater resolution) using regular single aperture radio telescopes round the world. The problem of measuring phase was solved by communications between telescopes to a central correlator. If the telescopes were too far apart to do this the signals were recorded on tape with precise timing with masers (atomic clocks) and then sent to a central facility where they could be processed. The Merlin array, based in Jodrell Bank, was an early example. Spread around the Midlands of the UK, the seven-telescope array has a resolution of about 40 milliarcseconds. Soon telescopes could be on different continents. With the increased baselines the resolution of these systems exceeded by some margin that of the best optical telescopes of the day.[25]

However aperture came with a significant cost – and not just the financial one. There was an intellectual challenge inherent in interferometry and aperture synthesis. As was said earlier, with an optical telescope all you have to do is place a photographic plate at the focus and you have an image.

With aperture synthesis you have first to gather the radio signals from all the elements of the array and then somehow process them to give you an image.

In the very early days this was done by hand calculation and took days. Computers came along in the late 50's and the computation were done much more quickly. Now it's possible to do them in real time.

The French mathematician Joseph Fourier, in the early years of the 18th century while studying heat conduction, discovered that pretty much any mathematical function could could be expressed as a series of of sine waves with different wavelengths and various phases. It could be extended to two dimensions so that for example a star field could be transformed into a series of frequencies and phases. This was just what an interferometer did.

To get back to the source field all that had to be done was reverse the process with the inverse Fourier transform. Aided by increasing computing power and a mathematical innovation known as the Fast Fourier Transform the calculations became much more tractable. While methods for speeding up the computations had been known since the early years of the 19th century, the FFT as a general tool is

25 Optical interferometers were to come along and change this

usually thought to have originated in the work of James Cooley and John Tukey in 1965.

The inversion was not straight forward because the response (the beam characteristics of the telescopes) had to be folded in and when it was actually done the results seemed rather disappointing not least because there seemed to be a lot of artefacts in the image that clearly did not reflect reality. Some appeared to be a result of the imperfections in the model of the array's beam structure (in reality it had side lobes and was "dirty") but others were a due to the sparsity of the collected data.

One response was to build larger and denser arrays with a good variety of baseline lengths (it had been recognised for some time that a mixture of lengths was a good thing). Another was to work out algorithms that made the best of the data that was available. As astro folk-law has it: in the USA astronomers predominantly followed the former approach while European ones tended to follow the latter. The truth may be more that European (especially UK) astronomy had made better use of computers earlier and had more faith in their powers than their American cousins.

The dominant way to clean up the images is the clean algorithm CLEAN.

This involves taking the actual results from the telescope and a function that represents the angular response of the instrument and convolving them (let's say merging). It's usual at this stage to give some weighting to the individual data points. In order to use the FFT the results have to be interpolated onto a grid in some way. The inverse Fourier Transform has to then be applied. As we said above, the results were a little disappointing so much so that the map produced is, officially, the "dirty map".

The clean algorithm presumes that we are looking for discrete sources again black background and starts with the most intense one in the dirty map, it subtracts a fraction with the shape of the dirty beam. It then does that again and again until it's told to stop. It puts this source into another image and the repeats the whole process for the next most intense in the dirty map until it's told to give up.

A final convolution with a clean beam gets you to a cleaned up version of the dirty image.

As Rohlfs and Wilson[26] wrote in 1999: "Whether this algorithm produces a realistic image, and how trustworthy the resulting point source model really is, are unanswered questions...".

Other versions have been developed more suited to extended sources

An alternative approach, the Maximum Entropy Method (MEM), has an extensive literature but is not much used in radio astronomy.

The essential problem is that it is not possible (or maybe not practicable) to have the complete data needed for an analysis and therefore assumptions have to be made to fill the gaps. For example, you may set missing data points to zero and that is done to obtain the so-called "principal solution". There are slightly more sophisticated assumptions such as weighting, mentioned earlier, or tapering to interpolate or extrapolate the data you have into the unknown regions.

The essence of MEM is that if you have a number of solutions (or in our case, images) you should choose the one that requires the fewest and weakest assumptions. To quote John:

> ... the function is chosen in a way that concedes greatest possible ignorance concerning that part of the autocorrelation function that was not estimated. This lack of dogmatism, embodied as a constructive principle, is sufficient to dictate a unique smooth extrapolation of the autocorrelation function beyond the measured range.
>
> The principle is that of being maximally non-committal about the non-available data. Formally this consists of maximizing the *entropy measure* of the spectrum.

The entropy measure to be used is still a matter of dispute.

John's contribution was three papers in the 1970's (Ponsonby JEB 1973),(Ponsonby JEB 1978),(Ponsonby JEB 1976). His 1973 paper looked at the possibility of developing an MEM approach that

26 "Tools of Radio Astronomy", Rohlfs K and Wilson T L, 2000, p213

applied to polarised emissions. It used some rather complicated mathematics that seemed to promise some solutions but left many unknowns to wrestle with. The two conference papers of 1976 and 1978 would have proved interesting. The 1978 one, entitled "The Problems of Maximum Entropy" would have been particularly so but, alas, all that remains is a very brief abstract that gives nothing away.

He seemed to revel in this slippery principle and the complex maths it involved but did nothing on it after 1978. Perhaps the problems, for John, lay in its essential vagueness and the accompanying risk of seeing what you wanted to see.

2D aperture synthesis and the Moon

Aperture synthesis radar (ASR) is used to obtain high resolution images of the earth by satellites. Remembering how interferometry gives us ways of combining signals from two radio telescopes at a large distance from one another to give resolution equivalent to a single telescope of diameter equal to that distance. Aperture synthesis radar uses the same principle. However we do not need two receivers. We can take advantage of the fact that the satellite is moving. So if we store the signal at some point and record it again when the satellite has moved on, we have two signals that we can process to get a higher resolution just as we do with the two fixed telescopes.

If we have a single earth-bound telescope and it's always looking at the same face of the moon it's difficult to see how the motion, essential to ASR, arises. It comes from lunar librations. These are the wobbles that the moon makes in its orbit around the earth. The man in the moon nods and shakes his head. Instead of seeing 50% of the moon's surface we see about 59%.

In 1968 Thomson and Ponsonby published a paper in the Proceedings of the Royal Society that applied some of the Venus work rather ingeniously to get a radar map of the Moon at 73 cm. It used the Mark 1 telescope using the fact that it was moving because of the moon's librations. The data gathered could be related to brightness (the radio brightness of course) through a Fourier transform.

The process worked and from one day's observations (14 July 1964) produced a map of the lunar disc. The resolution was not too good(about 3 arcsec)[27] so that craters could not be identified but it was adequate to indicate mountainous regions and distinguish them from the mares. Mountainous region were bright while mares had lower reflectivity.

John Thomson sadly died in August 1969 after working closely with John for several years. They seemed to get on and there was a mutual respect. Thomson appeared to have a marked influence on John's quixotic tendencies and his inclination not to bring things to a practical, engineered conclusion. His friends think that John's career suffered from the loss.[28]

The work was repeated with more data and reported in 1972. This time the resolution was thought to be about 1 arcsec. It confirmed the earlier results and found that there was now some indication of craters which stood out as brighter but still blurry spots.

The work at Jodrell seemed to peter out after that. Elsewhere, it went on and workers achieved resolutions of around 100m with terrestrial telescopes by the 2000s. The experience of men on the moon between 1969 and 1972 and the results from many remote vehicles gave very detailed information for specific sites. But earth-based radar gave an opportunity to generalise the data they produced to vast areas of the lunar surface. It was rather cost-effective too.

JOHN'S MASERS

The maser project started with an idea (probably Lovell's) that VLB interferometry could be used to track the Viking Mars Spacecraft which NASA intended to send to the planet in 1976. In an ambitious mission it was intended that a spacecraft would be put into orbit to photograph the surface and then a lander would be detached and land on the surface to make detailed observations. There were to be

27 The diameter of the moon is about 30 arcsec

28 John Hunter Thomas born in Sheffield on 3 May 1929 and died at 20 Chapel Close, Wilmslow on 12 August 1969

two separate missions each to be launched in 1976.

The whole system depended upon extremely accurate timekeeping and this required masers – colloquially, atomic clocks. The principles and practice of masers is probably outside the scope of this book but I have attempted to describe the elements of them in Appedix 2.

John started the maser project in 1973 as an integral part of the VLBI system. His expectation was that he would lead a team developing two phenomenally accurate atomic clocks. The team did not materialise and he ended up doing most of the work on his own.

Time was desperately short for such a project anyway but John set out with his plan to construct the best, most perfect, most accurate, atomic clock yet built. He may well have seen it as chance to demonstrate that he was not just an electronic engineer but someone who could cope with the challenge of a multi-disciplinary project under some pressure.

We do have some of the initial drawings from 1974 and he may (or may not – see below) have had a first working prototype in 1982. But this was of course around six years after the Viking mission had very successfully taken the first close-up photographs of the Martian surface revealing a rock-strewn lifeless environment.

As the immediate pressure faded, John was dumbfounded when Jodrell went out and bought a commercial maser. He continued to work on his own device and seems to have redefined the original purpose in claiming in his CV of 1992 that "This work started as 'public output' in the same spirit as my earlier work on the digital spectrometers."

It was some years before VLBI was established as a technique for tracking spacecraft. Its first use was in tracking weather balloons released into the atmosphere of Venus in 1985 and, notably, in following the progress of the Huygens probe as it landed on Saturn's moon, Titan, in 2005. Jodrell Bank would have been well ahead of the game in 1976.

The first progress report on his masers in John's files is from

September 1992 and there never seems to have been any external publication of anything related to this key project other than a few reports to international meetings – which have been lost. We have the CV of 1992, probably intended for internal Jodrell consumption. The only other information available is the result of an examination after his death of the two masers he built by Colin Hopper.

These sources suggest that we could summarize the situation by saying that neither of the masers worked satisfactorily. Since he worked on the project for over 20 years on and off we should ask what went wrong.

There are many engineering challenges in making a maser. Vacuums have to be maintained, a reliable source of atomic hydrogen in its upper energy states has to be established, a means of separating the hydrogen in these upper energy states has to be devised and a population inversion has to be created. The population has to be maintained in a storage bulb bottle and the bulb has to be kept in a resonant cavity so that stimulated emission can occur and be sustained. As John himself said in his CV "hydrogen masers are tricky transdisciplinary devices, and a great deal of what has to be understood to build them is not published."

However John seems to have had a first device that achieved "sustained oscillations" in May 1982. He recognised that this was just a step on the road to a practical stable time keeper and that much work remained to be done. The first maser was dismantled in 1992 for an inspection of residual magnetism and to replace its quartz storage bulb and fit new magnetic shields There was to be a discharge bulb of a new design, a revamp of the power supply and an improved version of the unit for temperature control.

He started work on the second maser sometime in the 1980s and after a myriad of engineering problems (including with the magnetic shields and trouble maintaining a vacuum) he reported in 1992 that he had got it "oscillating without difficulty". He postponed assessing the frequency stability (he had trouble with the monitoring equipment he had built) and instead directed his attention to the microKelvin cavity temperature control unit which was only in breadboard form and, anyway, there was scant evidence that it actually worked

In his first 1994 report he claimed that the re-assembled No. 1 maser was made to oscillate by "pure luck" after problems with the magnetic shields. He expressed his profound disappointment that little support had been forthcoming and particularly that the services of David Holland (a technician) would not be available to him in the future. It put his previous opinion that he could "see no reason why the masers should not be serviceable in some sense in a year's time" in jeopardy.

John prepared a second progress report in 1994.[29] It seems full of complaints about: glass-blowing facilities, the workshop, inadequate service from a printed circuit board maker, poor soldering practice by someone else, new problems with the microKelvin temperature controllers, careless subcontractors, the slow progress in constructing a computer interface and the general lack of support he had

In short, after 20 or so years since he began work, there still seemed to be a lot to do. In fact, just possibly more than he cared to admit.

It all hinged on whether he had achieved oscillations with or without injection of an external signal and whether the oscillations persisted after the external stimulus was turned off. Colin Hopper is convinced that John told him that the oscillations did not continue (at least for long) when the stimulus stopped so were not self-sustaining.

One cause of this behaviour would be that the resonant cavity was not resonating at quite the right frequency. The resonant frequency is determined by the size of the cavity and fine adjustments can be made by changing the temperature of the cavity – emphasising the importance of accurate temperature control. Even finer adjustments could have been made with the incorporation of a varactor (a small voltage-controlled variable capacitor) into the resonant circuit but John eschewed such things. John's temperature control system relied upon relatively small heaters and it would have been reasonable of

29 Progress reports ("What Help" 1994) and ("Research Committee Papers" 1994) in John's archive

him, if really sustained operation was not being achieved, to suspect it as a cause and turn his attention to it. Even at this stage, in John's own words "How well it keeps the cavity temperature constant it is too soon to tell."

However it had the effect of removing his focus from the actual problem of resonance to the analogue circuitry of the temperature controller. Whether he made any progress with sorting it out we have no idea.

There is one small clue that he did suspect that there was something seriously amiss with the resonant cavity. Colin Hopper found a baffle in the cavity that could be adjusted with small screws. It would have had the effect of changing the resonant frequency by quite a large factor but, because of the nature of the engineering, an essentially uncontrollable one. It may have represented a desperate attempt to find a solution to the resonance problem.

Other problems persisted. For example, the hydrogen supply was a continuous source of nuisance.

John made comments in his progress report to the Resources Committee of September 1992 that throw some light on the atmosphere at the time.

Firstly he explains that he had not submitted a report and not attended the committee's previous meeting in June because he had only been told about it while it was actually sitting. Secondly a report had been made to the meeting by "IM" without any prior consultation with him. He had not seen a copy and he thought the whole thing a serious breach of professional etiquette . He asked the committee to reject the report and expunge it from its records.

He then picks up an item from the minutes of the June meeting:

> "RDD (Rod Davies) sought the backing of the resource committee to discuss progress, development and publications arising from the maser work with JEBP."
>
> John commented:
> "I am at a loss to understand what this means."

He was probably the only person at Jodrell who didn't know what it meant.

There was an action against John's name about agreeing a performance specification for "the Jodrell maser". John advised the committee that "it is out of the question to start writing technical specifications for my masers at this stage of my project."

From this distance it seems incredible that he had been allowed to get so far in the project (20 years) before anybody asked for such a thing.

When he retired he asked if he could have his two masers and, when this was agreed, he had them taken away on a lorry and placed in his garage at Dean Close. He bought a lot of tools and a metal-working lathe with the clear intention of finishing the long-running job. He did try to find commercial storage for them elsewhere but no-one, understandably, seemed to care for his requirement that they had to be connected to the electricity mains and switched on at all times.

They briefly came to prominence in his ill-fated plan for the Scottish Centre when he acknowledged that they were "unfinished" but made them a central part of the offering.

It's not obvious whether he really did much to them: they stood as dusty stainless-steel monuments at the back of his garage, accessible only through a series of leaps over his other treasures. After John's death they were taken away for use as spares.

OTHER ELECTRONIC PROJECTS

Throughout his career John worked on numerous smaller projects, usually electronic in nature and often in support of major ones. They included:

an autocorrelator. This is described in detail in Appendix 3

an active all-pass filter using a differential operational amplifier

a sampling phase rotator for removal of Doppler shift in a lunar radar

quadrature hybrid made from transmission line of single

characteristic impedance

a simple Loran-C timing receiver for supporting VLBI

shapes of storage bulbs of hydrogen masers for maximum filling factors

hermetic sealing of a coaxial cable by a "double tourniquet"

SPECTRUM MANAGEMENT

Radio telescopes have incredible sensitivity, detecting tiny signals from space. We live in an electrical age with all sorts of sources of radio emissions at many frequencies both intended and spurious, at many frequencies. Grote Reber with his pioneering parabolic dish in the 1930's had trouble with spark plugs in car engines and radio-frequency interference (RFI). It has only got worse, much worse, since then. Telecommunication in all its aspects has grown explosively; the skies are alive with aircraft and, since the 1960s, satellites all communicating with someone or something.

There were many aspects involved in managing this situation and creating a radio environment where radio astronomy can function. Perhaps the most important of these (and the one John made a significant contribution to) is making sure that frequencies are allocated to all active users (those who produce the radio signals) that allow quiet bands for radio astronomy.

Terrestrial radio telescopes operate at frequencies that extend from about 15 MHz (wavelength about 20 meters) to an upper limit of around 600 Ghz (wavelength about 0.5 mm). This is because outside this window very little radio-frequency radiation gets through the atmosphere; it is absorbed by the ionosphere or by water and oxygen molecules in the atmosphere. There are some frequencies that are of particular interest to radio astronomers in this rather wide spectrum.

First attempts at managing spectra came when the ICSU (then the International Council for Scientific Unions but now the International Science Council) set up the Inter-Union Commission on the Allocation of Frequencies for Space Research and Radio Astronomy (UICAF). This had representatives of URSI (the International Union of Radio Sciences), the International Astronomical Union (IAU), and

COSPAR (the Committee on Space Research). These were the representatives of the telecommunications industry, radio astronomy and people with potential interests in other aspects of space research. IUCAF had a voice at the World Radio Conferences convened by the International Telecommunications Union (ITU) which had the power to create international law governing the use of the radio spectrum.

One of the key committees was the Joint Working Group of the IUCAF and Commission J (Radio Astronomy) of URSI called the Inter-Union Working Group on Radio Science Services.

In Europe the Committee on Radio Astronomy Frequencies (CRAF) – setup by the European Science Foundation – represented the interests of radio astronomers on frequency. John and Jim Cohen (also from Jodrell) jointly (and effectively) represented UK radio astronomy on several of these bodies.

The essential problem was that a pure sine wave carries no information. For it to do that it must be modulated or interrupted in some way and this inevitably introduces sidebands on either side of the pure transmission frequency. Similarly if the pure wave is distorted in any way it will produce sidebands too. The sidebands will spill over into adjacent bands and cause radio frequency interference.

Another annoying feature of electronic systems is their tendency to produce harmonics of the intended transmission frequency and these can be widely different in frequency.

Both of these effects can produce nightmares if you are trying to receive the monumentally weak signals that radio astronomers rely on.

There is a key frequency band important to radio astronomy between 1610.6 and 1613.8 MHz for study of the hydroxyl ion[30]. It is protected as a Primary resource for radio astronomy.

When they first launched in 1982 the Russian GLONASS satellites transmitted between 1602.0 and 1615.5 MHz, overlapping the radio astronomy allocation. After a world-wide rage in the astronomical community an agreement was reached by IUCAF and the Russians

---

30 This was a frequency range allocated to radio astronomy in the 1990s when John was active. The precise allocation at present is slightly different.'

to avoid frequencies in the band important to radio astronomy and to progressively over the next few years improve the filtering of the transmissions. In 1995, when Jim Cohen reviewed the situation, it was still difficult (Jim said "useless") to make observations of the highest sensitivity with single telescopes. Interferometry was less affected.

The American GPS system came on line for military purposes as early as 1973 and was fully operational with a full set of satellites in 1993. It was less of a threat to radio astronomy than GLONASS because the frequency it used (1575 MHz) was more remote from the radio astronomy allocation. However, it was noticed by radio astrnomers and the second generation of satellites were fitted with band-pass filters that virtually eliminated the problem without degrading the position fixing performance. It showed that things could be done at source to alleviate things.

The problems were much the greater because these global positioning systems were in lower orbits and, by their nature, had to have the necessary power to be received by small receivers on the ground.

A new threat in 1995 that greatly concerned John and Jim was the Iridium satellite phone system that Motorola (in the guise of a subsidiary Iridium LLC) planned to build that would interfere massively with the 1610.6 - 1613.8 MHz band (ironically, just as it was being cleaned up by GLONASS). This was originally planned to be a total of 77 satellites[31] in low earth orbit which would provide world-wide telephone communication. The down-link from the satellites to the hand-held phones was planned to be in just the region where there would be significant overlap with the reserved band.

The US NRAO signed a Memorandum of Understanding with Motorola on the basis of a time-arrangement and, allegedly, some kind of duress[32]. The time sharing would be achieved by Iridium

31 The atomic number of iridium is 77 so it was a nice choice. When they realised that they could manage with 66 satellites they kept the name. Atomic number 66 is dysprosium which would not sound as good by a long way.

32 The NRAO was/is a Government- funded organisation

sending a blanking signal to radio telescopes causing them to, effectively, turn off. Iridium would then transmit for a few tenths of a second during the blanking signal. At its end Iridium would stop transmitting and the radio telescopes would turn on again. The radio telescope would be turned off for 50% of the time.

This would mean that at least twice as long would be needed to gather equivalent data in the band and there would be knock-on effects on programmes that used other features.

It turned out that the NRAO would not be much affected since the band was not much used at their observatories. However other independent US labs refused to sign up as did the international community.

John could see no technical way out for Motorola because their proposal required amplifiers to be driven so hard that their outputs would necessarily be distorted and have out-of-band spurious emissions.

He seems to have been right about that because, under pressure from the EU, in 1998 the European Science Foundation, the umbrella organisation for CRAF, and Iridium signed a Framework Agreement. Coincidentally or otherwise, Iridium was switched on in Europe that year. An Interim Agreement was signed the following year which agreed quiet windows (well, maximum agreed interference levels) for European radio astronomy for seven hours every day and all day for two weekends a month The timings would be down to Iridium. The agreement was valid until 2006.

Iridium was finally fully launched in 1998. However it failed to attract many subscribers, partly because of the unexpected growth of terrestrial mobile services and partly because of the frequently poor performance in buildings. There were vast debts from all the launches too. After just a year the company filed for bankruptcy in one the largest cases in US history. However, radio astronomers could not really celebrate because the business carried on under another name after being bought for a paltry sum by another company. They planned to have a second generation system by the late 2000s.

John had retired by the time all this was sorted out. He probably thought, rather as Jim Cohen did, that, while it was a better deal than US astronomers had got with their Memorandum of Understanding, it was far from ideal. Jim Cohen could "see no direct scientific benefit for all the effort that had been put in".

Jim Cohen did continue the work on spectrum protection, becoming Chairman of CRAF from 2003 to 2006, but he sadly died aged 58 in 2006.

John must have felt some relief to have the episode over. According to his friends he had seriously considered his life to be under threat from the dark forces of capitalism as a result of his vigorous and public opposition to Motorola.

He mentioned "little accidents on the road perhaps" and "the kind of thing that makes one look under one's car before starting the engine" in his correspondence and gave his close friends instructions about what to do in the event of his death in suspicious circumstances.

Some friends found these right-wing conspiracy theories incredible and off putting and drifted away from him. Few really believed him. Most shrugged them off as just John being John.

RETIREMENT

When John learned from the Director of Jodrell Bank, Rod Davies, in January 1991 that his promotion had not been granted, he wrote to Rad saying that he intended to investigate early retirement on the grounds that it had to be financially better than straight resignation. However it seems to have been May 1997 before he took any real action.

He set out some of his reasons for eventually doing so in some hand-written notes from about this time:

> "... the whole University is no longer a place I feel proud to belong to."
>
> " I am thinking the thing is to apply for early retirement and get the hell out"

"Sick at heart"

" Everything I do or say or indeed fail to do or fail to say, displeases people"

"I feel stupid, inadequate, humiliated, chewed up and no good"

"I feel I just want to go off and disappear"

" I neglect my friends more & more. It's a bad business"

" I find myself in behaving in ways that I myself recognise as strange"

In a second note he complains about the quality of education, suggesting that the University presents material that students are insufficiently equipped to understand. Worse, they don't understand that really they don't understand. Not least because we are telling the world that they have achieved a standard that they clearly haven't. It approaches an academic fraud which brings the University and academics as a whole into disrepute.

In June he talked to Rod Davies, and to Bill Phillips, Director of Physical Labs. He was particularly concerned about continuity of the spectrum management work because Jim Cohen, who had been working with him on this, was about to be going off for a study year. John offered to fill in until Jim's return.

John had decided to proceed with retirement by early July and by 18 July the Early Retirement Committee had agreed that he should retire on 31 January 1998 with an enhancement of six years and 240 days, so what he would have had if he retired at 65.

As for the question of standing in for Jim Cohen, they offered him an appointment as a Fellow in Radio Astronomy to perform "frequency protection duties" for six months after retirement for a fixed fee of £1500.

John took a long time to reply to the offer . He seemed to accept it – at least the rewards package – but took strong exception to the university' plans for him covering for Jim Cohen.

In his letter to Andrew Lyne (who by this time had taken over as Director of Jodrell) of 6 November he described £1500 as "absurdly inadequate" and " derisory". He was expecting at least £4-5000.

John raised the matter in a final letter to Lyne on 23 January and commented:

"I suppose the truth is that the University sees no future for Jodrell Bank and consequently has no interest in the long term future of radio astronomy. It is too bad."

It was an extraordinary and likely offensive comment to make to Lyne. We can put it down to John's general despair but it wasn't a way to negotiate a better deal.

John was furious when he was told on 30 October that his discretionary points had been removed in accordance with a recommendation from the Faculty Review Committee. These were additions to his salary to recognise his work on masers and spectrum management and were worth around £3000 on top of his salary £28000.

John thought it a gross injustice and wrote to Bill Phillips and complained to Rod Davies. John was convinced that had the Committee understood his real contribution he would have continued to receive the payments. Phillips told him that he had asked department heads for their views and that anyway it was necessary to make cuts across the board because finances were not healthy. John believed that either Rod Davies had either not bothered to express his views or that he had submitted a non-supportive response because he did not understand what John actually did. The fact that it all would have no effect on his retirement package was little comfort. He still thought it was an injustice and an affront to his years of devoted service.

It was not the final injustice according to John. He was charged £81 for his car parking pass by deduction from his final salary payment. It turned out that he had not handed in his pass and was being charged for it until September – even though he was forbidden from parking after retirement.

Whether he actually handed in his pass and received a refund is not recorded but he did take the opportunity to raise a related matter: he had recently been charged £5.00 for replacement of a card that had stopped working. Careful reading of the university's car parking regulation revealed to him that this charge was to be levied on cards that were stolen or lost. There was no mention of defective cards so he told them that he would like a refund for that too.

There was one more thing to come to convince John that the university was incompetent and duplicitous. He was astonished in February to find that he had been paid the first instalment of the derisory and rejected £1500. He wrote a letter of complaint. Whether he sent the money back is not recorded.

The final act was an agreement that he could take his two masers with him. They were delivered to his garage on a lorry.

5: FRIENDS

John had many friends from diverse backgrounds, too many to detail properly. Several have been mentioned in the text but there was a close, special group, a lady in Sweden and an Indian engineer who in some sense stood for all the others

THE GROUP

John kept in close touch with one group of friends he made when he first went to Manchester to do his PhD: Simon Lavington, Rosalind Twyman (later to marry Simon), Graham McLean (always known as "Mac"). Sue Littler, later Sue McLean, also became part of the group.

John met Simon and Rosalind at the University Christian Union. Mac and Sue met them at a party at their house in Didsbury. Ann Matthias was a friend (they were in the same hall of residence) of someone Sue had been at school with. The connection was made when Ann was asked by the head of the hall of residence to keep an eye on Sue's friend, who was alarmingly distressed by the death of her father and mother in a road accident: she had been the sole survivor of the crash.

Simon was doing a PhD in computer science, Rosalind was in the middle of an undergraduate course in history, Mac was doing a PhD in electrical engineering and Sue was training to be a physiotherapist.

This diverse group, brought together by disparate routes, was to stay close friends until John's death. They went on holiday together, stayed with one another, met up for birthdays and other family events. As children came along, they became friends too.

Most Sundays John had lunch with the McLeans.

THE GROOVE[33]

John formed many friends amongst the research students at Manchester University in the 1960s, particularly those associated with the Electrical Engineering Department in Dover Street wherein resided the powerful Ferranti Atlas supercomputer. Use of this, and other much smaller computers at Jodrell, became part of John's skill set. John remained in contact with this so-called *Groove* group of Manchester friends for the rest of his life, appearing regularly at *Groove* parties.

Mac was a leading light of the *Groove*.

Illustration 12: Mac and John with Mac's glider

Illustration 11: Rosalind and Simon

RENATE

John met Renate Nyman by chance. On a visit to France the window

33 Named because many of them lived in a road called Grove Avenue

of his car was smashed and his camera was stolen. Presumably the window was fixed quite quickly but the camera was not replaced straight away. It is likely that John was on his year of secondment to the Onsala Observatory because he chose to go into a camera shop in Kungsbacka, a town around 30 km from the Observatory, to buy a replacement. There was Renate. She lived in Kungsbacka.

She was a young (nine years younger than John) single mother, German, with two children living in Sweden. Whether she was divorced or estranged we have no idea. She was trying to make her way as a professional photographer. Other than that she wrote to John most days between July 1996 (John returned to the UK, his secondment ending, on about 19 September 1996) and November 2002 and compiled a letter to send about every week. The hand-written letters totalled about 400 000 words and fill seven ring folders.

John's response, by email, is lost. One letter survives.

The correspondence was lively with Renate reporting on family matters, books she was reading and, occasionally, God and religion. Dipping into them we find she had diverse interests. The works of Lewis Carroll (she referred to him as "MrD"), particularly the enigmatic *The Hunting of the Snark,* which seemed to be much in her mind most of the time.

She was also intrigued by Elaine Morgan. Morgan, a journalist turned evolutionist, who propounded a theory that man evolved not as a hunter on Africa's arid plains (a popular notion still) but as a gentle, aquatic creature of the coast, more gatherer than hunter. It wasn't actually her theory but had been proposed by Alistair Hardy in 1962. However she wrote books on "The Descent of Woman" and "The Descent of the Child" based on it, seeing that it gave women a role equal to that of men in the evolutionary process. The books were scorned by the professionals but were massive best sellers.

Renate's letters were a well-written, eclectic stream of ideas and imaginings in a bold hand by a clearly intelligent woman. The only theme that really emerges is her love for and infatuation with John. She addressed him as "dearest", "dear love", "dearest Happy Thought", "Big Boy" and "My little Space carrot". Fairly often it expanded into asking him if he felt the same about her.

Of course we don't know what his answer was (which is a shame) but, since she kept asking it for nearly six years, we must assume that John didn't respond in the unequivocal affirmative. He probably hedged. Or maybe he did what he did so often in these situations and replied with another question. Or maybe he started a general rather philosophical discussion of the nature of love. Whatever, the correspondence drew on, Renate stopped asking but until its end she regularly avowed her love for him, requited or not.

The letter that survives – well part of it does – is rather revealing because he talks about his relationship with his mother.

The correspondence we have ended in November of 2002 and the relationship shortly after that. It was probably ended by John.

Renate was diagnosed with cancer and she died on 29 April 2013.

IQBAL AND A VISIT TO INDIA

On the 23 December 2002 flew out of Manchester to go to a wedding in Bangalore via Heathrow, Bahrain and Bombay – Business Class of course – and then by limo to Pune(Poona). He stayed in Pune on Christmas Day and Boxing Day and then flew on to Bangalore on the 27th. His luggage included a sleeping bag because "Indian bedding leaves much to be desired".

The wedding was of Vivek Radhakrishnan and Namrata Kini. Vivek was the son of Professor V Radhakrishnan (John called him 'Rad') and Dominique. Rad was the Director of the Raman Research Institute in C V Raman Avenue named after his father Sir C V Raman, the first Asian to be awarded, in 1930, a Nobel Prize for science.(Rad's cousin Subrahmanyan Chandrasekar was also awarded a Nobel Prize in 1983 for his work on stellar evolution.)

Dominique was a Parisian whose real name was Françoise. However Indians had such a problem with pronouncing it that she was known as Dominique

Vivek had studied design in Eindhoven in the Netherlands and was, at the time of the wedding, living in New York. There was something so international and high achieving about the family that no wonder John felt at home.

The celebrations lasted three days and on the 31st John returned to Pune and then to New Delhi to visit friends at the National Physical Laboratory. After this, on 5th January, he flew to Nagpur to meet Iqbal Ahmed.

Iqbal was a model-maker who built exquisite, miniature working-models of things mechanical: trains, steam engines and cars mainly. Perhaps John knew him through his achievement in building the smallest working steam engine in the world: it could sit on a fingernail. Anyway, John had been in correspondence with him for a year and had looked forward to this opportunity to visit him and see his work – including the mighty, small engine. But first he wanted to do something else. He wanted to find the grave of another of his heroes: Colonel William Lambton. He and Iqbal, with, Iqbal's son and daughter, set off in a borrowed car to Hinganghat about 50 miles south of Nagpur using directions provided to him by John Keay, the author of a book about the Great Trigonometrical Survey of India.

The Great Survey

The Survey was proposed by Lambton, an experienced surveyor, and he was appointed Superintendent when it was agreed.

The Survey was later extended to cover the whole of mainland India but it began with Lambton accurately measuring the length of one degree of latitude in April 1802. This would establish the shape of the Earth, assumed to be a flattened sphere (an oblate ellipsoid of revolution to be precise). An accurately-known baseline could then be laid out and the process of triangulation could be used to build up a triangular grid extending across the country. The triangulation point had to be in sight of at least two others so it would be located on high ground or a tall building. Occasionally towers had to be built. The Great Theodolite used for the angular measurements therefore had to be carried by porters (roads were too rough for it to be transported by bullock cart). It weighed around 500kg.

The Great Theodolite suffered a terrible accident at Thanjavur in late 1808. While trying to hoist the half-tonne theodolite to the top of 217-ft gate tower of the Brihadeeshwarar temple the instrument crashed down into a tangled mass of steel.

Rather than wait years for a replacement, Lambton shut himself up in a survey tent for nearly six weeks and repaired it himself with some help from a nearby military workshop. It was more or less as good as new. And it was, one feels, exactly what John would have done in the circumstances.

Lambton continued with the Survey extending it from east to west and the north to south across the country until in 1823, aged 70, he passed away (probably from tuberculosis) in a survey tent near Hinganghat and was buried there. The grave had been described by John Keay in his book and John managed to get fairly detailed instructions to locate it from him. With local help he found it quite quickly: located on a desolate waste ground close to a mysterious black column known to the locals as Kala Gota and by Keay as “the Belgian’s Stone”. The grave was a slab in a bad state of repair with fragments of roughly-drawn writing (as if by a finger in wet cement). On the first line Keay could see the letters LAMB when he was there; John’s party could see only RAM. Below it was written:

*ORDE

DATE*1766

The asterisks represent different signs that can’t be interpreted. John’s photograph of the inscription is below.

Illustration 13: The inscription

Keay was puzzled by the date (he seems to have been able to read only the first three digits 176). Anyway, obviously not his date of death, it might have been his date of birth. This is uncertain but an accepted value is 1753 (Gleanings In Science 2 1830). So 17 sixty something did not seem impossible.

John pointed out the problem with the name (but not the date) when he wrote to Keay after his visit. Keay's response is not recorded.

After an inspection of the Kala Gota where no inscription could be found (John was deterred from investigating thoroughly by the strong smell of urine around the plinth) and after taking a GPS reading and realising that Iqbal had had enough, they left for Nagpur.

John had taken care to record the GPS position of the grave at 20°23.12'N and 78° 50.42' E. This, according to Google Earth, lies in an Urban Primary Health Centre. The Kala Gota and hence the grave lie about 300m to the north east close to a spot labelled "The Lambton Cemetery". It is also nearby the position labelled "The Centre of India" and the location of one of the GTS triangulation points. So it is good that Lambton lies so close to the centre of his universe but slightly ironic that we cannot be absolutely sure that he is down there and, if he is, John's GPS or Google was a bit suspect. Lambton you feel would not have been amused.

A 2015 photograph of the Kala Gota shows it has been given a new plinth, surrounded with a concrete and metal fence and designated as the Lambton Memorial. The grave is within the fence.

Nagpur

They arrived back at Nagpur for a late lunch in a restaurant run by one of Iqbal's friends and the went to Iqbal's house which he shared with his two brothers and their families.

Iqbal was a self-taught engineer and made a living from his lathe making spare parts for anything from photo-processing machines to cars. He was making a replacement camshaft for a Peugeot car when

John visited. The image shows Iqbal and his lathe in his workshop.

Illustration 14: Iqbal's workshop

A supremely absorbed craftsman surrounded by apparent chaos. John probably felt at home with that.

John spent the afternoon with Iqbal looking at his models and inspecting the smallest one, which normally sat demurely on a tiny cushion. He came back the following day too so fascinated was he. And the next day he set off by car, back to Pune.

Leaving at 0840 he stopped at Garanja for lunch, at Lonar for the crater and arrived at his overnight stop, Aurangabad, at 2000 after a 455 km drive.

He allowed himself half an hour to look at the spectacular Lonar Crater, created by the extremely high- velocity impact of a comet or asteroid some 600,000 years ago The crater is about 1.8 km wide and contains a lake which has been known to change colour. There are several notable temples around the lake.

He stayed Aurangad for 3 nights,,visiting the Ajanta Caves (a feast of carvings and rock paintings) and then continued his drive to Pune on the 11th.

He visited the Giant Metre-wave Radio Telescope (GMRT) on the 15th. The magnificent facility is composed of thirty 45 metre totally-steerable dishes, about half of them in a central cluster and the others in a great Y-shape extending across the countryside for 15 km in the three direction. Started in 1984 and finished in 1986, it is an interferometric array of international importance withbaselines of up to 25 kilometres.

John returned to Pune on the 16th visited MIT on 18th and on the 20th travelled to Mumbai for a flight home on the 22nd.

6: SAILING

John bought his first cruising boat in 1971. She was a Dragonfly, an 18ft GRP vessel designed by Alan Buchanan, called *Phaethusa.* She had three berths and was just five years old, smallish but well equipped for off-shore cruising. John kept her at Beaumaris on Anglesey.

His first recorded sail was from Beaumaris to Holyhead on 13th April 1971 and back the following day. The winds were fairly light so it must have been a pleasant and encouraging start. By 29th May he was ready to try something adventurous: he set off at 1 am, intending (it says in the logbook) to go to Port St Mary in the Isle of Man. For some reason at 5 am, after he had gone 10M and was to the north of Anglesey, the logbook stops.

There were other entries for 1971 but they simply record the weather forecast. Even they stopped on the 31st of May. There is a single entry for 14th July 1974 of a not-very-good forecast and then nothing until 14th September 1976

The truth may be that he had realised that he needed a bigger boat to do what he wanted to do – even as he sat in the dawn light off Anglesey in 1971.

Eventually he did act. On 1st December 1975 he bought a bigger boat, *VrelliII*, a Trident 25 based at Emsworth. He put *Phaethusa* up for sale in April, fixed *VrelliII* up over the winter, spring and summer and on 14 September 1976 began to bring her north. He went via Lulworth Cove, Brixham, Salcombe, the Helford River and Newlyn (a 10 hour passage that made him write "thank goodness!" in the log). After a couple of days of rest he set off again on 25 September at

10 am to arrive at Dale on the Pembroke peninsula by around 9 am the following day, a distance of just over 100M. He set off again on 28 September at 10:45 am, and after an overnight passage of just about 100M, arrived in Beaumaris at 6:20 am.

Illustration 15: VrellII at anchor

The reader will be asking by now if he did all this his own. The answer is " possibly". John was not in the habit of recording if he had crew and he certainly made no mention of anyone in this cruise. The log is filled out hourly in his own handwriting but given John's perfectionist nature he may have insisted on doing this. Whatever, the voyage was a remarkable achievement in his new boat and it must have given him confidence .

According to the logbook he had had his adventure for the year and he left her at Beaumaris for the winter and didn't sail her again until 30th July the following year when, in a rare moment of poetic reflection, he wrote in the logbook:

Puffin Sound. The spinner is in the water. The sea is flat. The wind blows from the north. The sky is low.

By the end of the day he had reached Port St Mary on the Isle on Man. The sea was still flat but he had needed to motor into a force 6 to get there.

By the following day he was in Portpatrick and by the 2nd August he reached Sanda and from there went to Gigha, then to Gabhard Island, to loch Drumbuie by the 6th. At his point he turned around and headed south back to Gigha, Portpatrick, Port St Mary and finally Beaumaris on the evening of the 14th.

He went back a week or so later to have a run to Caernafon for the night and then on to Eastham Lock and up the river Weaver to her winter home at Jalsea Marine in Northwich.

Alas we do not know what happened in 1977 to 1979. Either the logbook has been lost or John kept her for two years at Jalsea Marine to do work.

The next logbook we have starts on 28th July 1980 and a journey north. For once we know the crew was "Anders the Geek". The route took him via the Isle of Man and Port Patrick to Ardrishaig on the 4th. The crew changed: Anders left to be replaced by Peter Wilkinson.

Peter sailed with him as far as Mallaig on the 9th. By the evening of the 10th John was anchored off the island of Rhona to the east of Skye and on the next passage to Greshonish he reached his farthest north; by the 14th he had completed a circumnavigation of Skye. He made his way back, diverting to Tobermory, to arrive in Liverpool on the 25th August after completing a 600M voyage, largely on his own – probably.

In 1981 he went round to the south coast of Ireland as far as Bantry Bay on the extreme south-westerly tip. Starting on the 29th of July he was back at Menai on the 18th of August, again having covered around 600M.

He did take a trip to the IOM over the Bank Holiday weekend at the end of August but he was back at Eastham Lock at the end of September for his winter layup.

There is nothing in the logbook for 1983 but in 1984 he went north again in almost a repeat of his 1981 voyage.

There is a break again in the logbook for 1985 and 1986 but for 1987 he had something special in mind: he would go to the Faeroe Islands. He planned to take Russell, the young son of a Professor at Onsala with him. Among John's papers is a letter from the Professor formally giving power of *loco parentis* to him for the adventure.

For some reason it did not start well. He set off from his berth and went to the Boat Museum on the 5th August and then stayed there until the 9th – just possibly waiting for Russell. The weather looks as if it was OK. He was in Portpatrick by the 13th, in the Kyle of Lochalsh by the 19th. That night they were alongside a fishing boat in Loch Gair aiming to go to Stornoway.

Come the 20th John bought some paraffin in the village, put Russell on the Inverness bus and set sail. He was abeam Longa Island in the entrance to the Loch when he got a strong wind warning from Oban Coastguard of S becoming SE with force 6 at times. He decided to abort the passage to Stornaway and turn back south. He was slightly more north than he had been in 1980 but still about 250 M short of the Faroes.

Maybe rather deflated, he retraced his steps south until he moored in the Mersey at 1:40 am on the 2nd September. By teatime he was back at his berth in Northwich.

Nothing happened in 1988 but in 1989 he was off north again. He left his winter berth on the 17 th of July and by the 22th he was at Rum. The next day he sailed to Loch Maddy on the Isle of Benbecula (a 50M passage) in the Outer Hebrides. By the 25th he was in Stornoway (at last), he was stormbound in nearby Loch Grimshader for two days and then, on 28 July, he struck out for the mainland. By that evening he was in Loch Inchard, just 10M south of the north-west tip of the Scottish mainland. This was his furthest north ever; just 200M south of the Faeroes.

After a one-night stay he retreated south and was back in Tobermory by the 4th August. He got to Ramsey on the 13th and on the 15th he heard that his mother had died. He felt that had no option but to carry on and so he did and by the 18th he was back at Northwich.

In 1992 he went up to Scotland yet again reaching just short of

Cape Wrath. He was back at Liverpool and his final entry in the log was at 1350 on Monday 31st of August "tied up against a concrete barge at Boat Museum." He then must have taken *VrelliII* to Jalsea marina for the last time: his active sailing career was over.

VrelliII never moved again, other than being shuffled about on the yard, until six months after he died. That was on a trailer to the Norfolk Broads .

Quite why he gave up is a mystery although a clue emerged from the papers he left behind. Although he seems to have given up in 1992 in September 1993 he took a practical exam for the RYA Yachtmaster – and failed. He was awarded the (lower) Coastal Skipper certificate instead.

The major deficiency seemed to be of his "general boat handling with a full crew". In a comment on his overall ability the examiner said "Dr Ponsonby does not have much experience of crew management, no serious awareness of team leadership".

Maybe John saw a grain of truth in this and saw that it was an essential missing ingredient if he wanted to continue enjoying his sailing. Maybe he was so affronted that he decided that he'd had enough and just gave up. Maybe – and this is possible – he didn't understand what the examiner was on about, dismissed him as a fool and the RYA as a corrupt, incompetent organisation and cashed his chips in.

The whole thing was somewhat ironic. It was noted earlier that he seldom recorded that a crew was present, even less who they were. It was as if they didn't exist. Maybe for John they didn't.

Whatever, *VrelliII* remained an important part of his life for a further 29 years after that final cruise. He was frequently on the boatyard making improvements and maintaining her until late in the evening and night. Whether he thought he would ever sail her again is doubtful and part of the reason for his continued interest was that he enjoyed the company of the long-term residents. It was a quiet little community of men that welcomed John with reminiscences and cups of tea. They called him "the professor". He liked that.

8: POLITICS

He was a paid up Liberal Democrat (as he told their Leader, Paddy Ashdown MP in a letter of complaint in 1990) and a Liberal before that. What he certainly was, throughout his adult life, was a supporter of peace movements. In his will he left a considerable sum to Greenpeace. But long before that, in 1964, he supported the British Council for Peace in Vietnam, chaired by Fenner Brockway, and, probably, Bertrand Russell's International War Crimes Tribunal, which found the USA and its allies guilty of war crimes in Vietnam.[34]

The decision by the Government in 1981 to allow US nuclear cruise missiles to be based at Greenham Common in Berkshire triggered the Greenham Women's protest that were to go on for a decade. It also reinvigorated the nuclear disarmament movement. It led to the setting up of the Manchester University Disarmament Group by academics and, internationally, Scientist Against Nuclear Arms (SANA). John joined both of these as well as, it is thought, forming links with the Bulletin of the Atomic Scientists.

John was not afraid to discuss religion or politics at length. He had strong principles. He made regular donations to charities such as Greenpeace, Campaign Against the Arms Trade, CND, Liberty (formerly National Council for Civil Liberties) and Anti-Slavery International. He once confessed that "For years I was mighty annoyed with my cousin Francis Ponsonby, who was Commander RN and at one time commanded a nuclear sub complete with Polaris missiles".

---

34 He was sent a letter signed by, among others, Jean-Paul Sartre and Isaac Deutscher seeking support for the Tribunal.

9: CAMPAIGNS AND COMPLAINTS

John was involved in many campaigns and seemed to relish complaining. His immaculate but discursive prose is a joy to read as a neutral.

KAL 007

The Boeing 747 of Korean Airlines flight 007 took off from Anchorage in Alaska at about 1200 TC on 31 August 1983 bound for Seoul in South Korea. Due to faulty navigation it deviated from a route that would keep it clear of Soviet airspace and passed over the Kamchatka Peninsula and then, some time later, the southernmost tip of Sakhulin Island both of which were extremely militarily-sensitive areas.

The Russians attempted to intercept the plane as it went over Kamchatka but failed. However, as it went over the island and thinking it to be a threat, they shot it down with a missile from an interceptor jet just as it was leaving Soviet airspace on the island's western coast. All 269 people on board died as it crashed into the sea.

It has taken some time to work out what probably happened. First why the aircraft was so far off the usual route (by the time it was shot down it was around 200 miles north of where it should have been) and why the aircrew did not notice the deviation. This was a long flight over open water after you left the Alaska mainland behind and it had to be made by dead reckoning using an inertial navigation system. It depended on very much on setting the system up on departure. If this was done properly then the navigation system would follow a preplanned route within just a few miles. A quirk in the system was that if the aircraft was not within 7.5 miles of the preplanned one it would take it on a slightly different course. Crucially the system gave no indication that it had done this.

This seems to have been what happened on KAL007 because subsequent analysis showed that a slightly incorrect course had been set right from the start.

There were quite a few opportunities for the crew to have identified that something was not quite right. Indeed they must have reported reaching waypoints along the route without any real checks. True there were a few radio problems and messages had to be passed by another KAL aircraft. True that the crew had had minimal rest between flights. But also true that the flight was notorious among KAL for being uneventful and boring yet stressful because it was over such a long stretch of water.

President Ronald Reagan described the shooting down as a "massacre" but the truth was it was a case of mistaken identity and incompetence. The Soviets thought that they had an intrusion by a US military surveillance aircraft on their hands and followed their procedures for that: the intercepting aircraft wiggled its wings as a signal for the Korean jet to follow and it fired warning shots. The warning shots were not seen because they were not tracer bullets and so were invisible in the dark. The airliner started to climb as part of an air traffic control procedure but the Russian pilot interpreted it as an evasive manoeuvre. Somehow it was falsely reported that the airliner was flying without lights.

When the radar data was analysed later it was clear that the fighter was never less than 7 km from the Korean plane. From that distance at night it was not possible to make a positive identification; even the distinctive bubble on the top of the 747 was not visible because the fighter was following at a lower altitude. No-one considered calling the airliner on the radio.

The people in charge of the interception must have been aware of two things: that the Korean plane would shortly be back in international airspace and that the Soviet procedure for handling intrusions demanded that, after reasonable steps had been taken to resolve the situation, the offending plane *must* be shot down. They gave the order to do so and the pilot of the interceptor launched two missiles. On missed but the other did enough damage to depressurise the Korean plane and make it unflyable. It spiralled down and plunged into the sea.

The aircrew of KAL007 were completely unaware that anything was amiss until the missile struck.

I have described the official explanation for what happened. But this was not accepted by a significant and vocal number of people and alternative explanations emerged. A few of these were:

> It was a planned spying mission that went wrong. The crew of KAL007 collaborated
>
> It was a diversionary tactic, that went wrong, to cover other US surveillance intrusions
>
> There had been a large air battle, probably a result of an incursion, with several losses on both sides, including the airliner
>
> The airliner had been intercepted and had landed in the USSR. The crew and passengers were then taken hostage.
>
> The crew were tricked into making the incursion because the Soviets had created an array of false navigation beacons. They could then justify shooting the plane down. They did this because an anti-communist Congressman was on board.

Quite how John became involved in this community we don't know. We don't know either what contribution he made although it seems to have arisen through his knowledge of satellites, which were thought by some to have been involved. He claimed that his involvement had been "minimal" but he did say "there was something very fishy about the whole affair" on the basis that the US government was not conducting a thorough and open investigation into what happened.

He regarded his interest in "the matter [the shooting down] as a closely guarded secret which I have been been eager to keep from officialdom."

So it was a shock to receive a newsletter, "THE KAL007 Information Bulletin and Newsletter" which named him as contributing his expertise to the satellite connection. Worse still in, John's mind it came in an envelope emblazoned with "THE KAL007 Information Bulletin and Newsletter" closely followed by another bulletin that gave his name and address (Jodrell Bank).

He called it a monstrous liberty that might not only compromise his ability to get into the US in future but might put his job at risk. Also, darkly, "There have been warnings that 'accidents' may befall people who ask too many questions."

THE AFFAIR OF THE NATWEST BANK

In John's words:

> Manager
> NatWest Bank plc, Wilmslow,
> 4,Grove Street, Cheshire,
> Wilmslow, SK9 ZER
> SK9 lEJ
> 27 June 2019
> Dear Mrs xxx
>
> I came to your branch on Monday 24th June afternoon in order to arrange a simple electronic transfer of funds to the account of my good friend xxxxx xxxxx.
>
> I had expected this to be a quick and easy operation.
>
> Your colleague at one of your counters viewed my passport and swiftly completed the transaction and had already handed me the Payment Confirmation slip when you intervened and reversed the transaction and created a nightmare of difficulty.
>
> I do not know what provoked you to do that. Not only did

you reverse the transaction but you called a fiercely armed police-lady to come to your branch. What you expected her to do I have no idea. When I first encountered her I greeted her in a friendly manner and had no idea you had called her to talk to me. When I found out I was thoroughly alarmed.

On Monday evening I dialled xxxxxxxx, which seems to be the only NatWest number available to the public, and after speaking to someone I was told I must speak to someone else on the SCAM's Team. After a very long, and doubtless expensive, wait my call was transferred to a guy who said his name was xxxx. He advised me to return to your branch the following morning and he said he could see no reason why you had treated me the way you did. He said I would be able, if necessary, to get back to him on xxxxxxx.

So I came back to you on Tuesday morning.

Despite my asking several times you gave me no clear explanation as to why you had acted the way you had and you said very firmly that you were not going to reverse your decision. Also you said I would not be able to transact my business at any other branch of the NatWest. This statement was in precise contradiction to the assurance that xxxx had given me on the phone.

You made me sign a paper which inter alia stated that "I/We have read and agreed to be bound by the Terms and Conditions" etc... even though I had not been shown these Terms and Conditions and have still not seen them. You then tried to make me go away with the promise that someone somewhere, on the basis of I know not what considerations, would phone me later in the afternoon to deliver their VERDICT. This was an unwelcome suggestion which I declined to follow and I said that I would prefer to wait with you until the matter was cleared up. At this point you again threatened me with involvement of the Police.

In the event the message came surprisingly quickly that in fact all objections to my transaction were lifted and the transaction then went through smoothly enough, despite your colleague disappearing behind the scenes and keeping me waiting for a remarkably long and unaccounted for time.

Yesterday I had an email from Mr xxxxx to tell me that the monies have satisfactorily reached his account. So after these vicissitudes the end result is satisfactory.

Strangely shortly after I got home I received a phone call from Wilmslow Police. They asked me if I wanted them to come round to see me. I told the young lady with whom I spoke that whilst it would of course be a pleasure to meet her I didn't think there was anything more to discuss and she didn't press the point. She gave me her name and service number and suggested that I might care to drop in sometime when she is again on duty to make her acquaintance.

....

I am yours sincerely
Dr John E.B.Ponsonby

THE JACKET

xxx and xxxx
xxx Regent Street
London

31st July 1971

Gentlemen,

I recently purchased a tropical suit from you of which the jacket is herewith enclosed. See your invoice no xxx.

This jacket is unfortunately not a very satisfactory fit. It is fine when I stand straight, but is disagreeable when sitting, especially when driving or sitting at a table.

The trouble is that it becomes too tight under the arms, across the shoulder blades and round the biceps. I have put in a number of pins round the left shoulder at the points where it pulls tight, and they lie along the lines of tension. I hope they are sufficient to indicate the trouble.

Please could you alter it for me by letting out as much as possible within the limitations of the material available. It will be convenient if you could than return it me , c/o University of Manchester, Jodrell Bank, Macclesfield, Cheshire

Yours faithfully
Dr J E B Ponsonby

THE LEAP SECOND

Leap seconds are occasionally added to clocks, which run on Coordinated Universal Time(UTC) – and are so determined by an atomic frequency – so as to keep them in step with the rotation of the earth. The rotation rate of the Earth is slowing down because mainly because of tidal forces but there are other smaller effects too.

Leap seconds were introduced in 1972 at the same time as UTC. Since 1980 eighteen leap seconds have been added in either at the end June or December. The decision whether to add one is taken by a committee of the International Telecommunications Union.

There had been a few problems with managing the leap seconds:

since the decision about whether one was to be inserted was made just a few months before it had to be done, computers could not be programmed to insert it, some computers and the software on them did not take well to odd seconds being inserted, and testing systems for compliance was difficult.

On the other hand the system had been operating since 1972 without major practical issues. It was much appreciated by astronomers, keeping their telescopes pointing in the right direction.

So it was a surprise to many when, in 2005, the USA proposed, through an ITU expert group, that the ITU should abandon the leap second. The main opposition to the proposal came from Canada, China, Germany and the UK. It was decided that a decision would be made at the next ITU Radiocommunication Assembly in January 2012.

At the Assembly it was decided to defer a decision until the next Assembly in 2015 to allow further discussions. When 2015 came the decision was to carry on using the leap second and to review the matter at its Assembly in 2023.

In 2023 the ITU followed the proposal of the General Conference of Weights and Measures that the leap second should be dropped by or in 2035 and replaced by, perhaps, the leap minute. A leap minute would be required about once a century. Plenty of time for computers, programmers and even astronomers to figure out how to do it properly; a welcome diplomatic procrastination.

Of course, by 2023 John has passed. But he did leave his views on record in 2014 in a letter to Stuart Carruthers, then the Cruising Manager of the Royal Yachting Association, and after after a crystal-clear summary of the issues (including mention of the Julian and Gregorian calender) :

"In my view abolition of the Leap Second is a deplorable and retrograde proposal that needs to be vigorously opposed."

THE MISSING REFERENCES AND OTHER THINGS

Monday 28th May 2018
Natwest Bank
Dear xxxx

Thank you for phoning me the other day. I found it difficult to follow everything you said. That wasn't your fault but is a peculiarity of my telephone answering machine. At the time I bought it, it was I thought the top of the range, but it has always been difficult to make out what people say, particularly recorded messages, unless they very deliberately speak up. The lesson I learnt was never again to buy anything from Samsung.

What I understood was that for some reason you thought you had made some mistake in the first version of your letter and so you were going to send again. I have received both versions and don't see any substantive difference between them except in the number you call "Your Reference:". Actually neither of these numbers originate with myself but are in fact your reference numbers.

They are PHO-0193366218 and PHO-0194674818 and neither agree with your reference number 18346862 given on your letter of 10th May and which for some inexplicable reason I managed not to see. I must apologise for my oversight.

I understood you to say on the phone that you were going to compensate me for what I expect to be a huge telephone bill by crediting my bank account with £25. That is very good of you and I thank you. It will of course be several months before I next receive a bill from the telephone people, and only then will I discover exactly how much the long telephone call will have cost.

I also understood you to say that you would be sending me

a Hamper. I could hardly believe what I thought you had said. But I had understood you correctly and indeed it arrived on Saturday. It is full of goodies and I thank you very much for it. Most of the items are sweet. Since I'm trying to cut back on sugar it will either last me a long time or I will pass them on to someone.

You repeated mention of these gifts in your letter.

I am rather shocked by your explanation of the illegible signature on your earlier letter. You say it isn't in fact anyone's signature. So it is a total fraud. Banks of all institutions surely consider signatures as reliable authentifiers and frequently and very properly require instructions to be in signed written form. You should not be abusing the signature system. Also it is usual for signatures to be accompanied by the name clearly written and for the very reason that it makes it possible to address a reply to the right person. I had this very difficulty with your mock signature.

The story about the old cheque is this. Long ago a distant relative of mine, of whom I'm very fond, was seriously swindled by a business partner as a result of which he was stranded penniless in the Caribbean without the means to get back to England. The Foreign Office got him home which left him with a debt to the FO. Until that debt was paid off he couldn't get a new Passport. In a fit of generosity I said I would pay the debt for him and the cheque about which I made enquiries was made out to the Foreign Office for the full sum owing. Since it was a crossed cheque that was certainly cashed I have no doubt that it reached the Foreign Office. Unfortunately I have no official documentary evidence that it reached the Foreign Office. I don't keep bank statements for more than eight years and all I have is the entry in my old account book. My relative also cannot find any old receipt but he now wants a Passport and the Foreign

Office are now saying he still owes them his repatriation debt. They too don't seem to have a record of the payment. So I phoned the NatWest on the off chance that just possibly you might be able to produce a record of the cashing of the cheque. I didn't seriously expect you to have a record but I did think it was at least worth asking.

There is another small matter to which I'd like to draw your attention. On Tuesday 5th May I had an appointment to see someone at the Wilmslow Branch of the NatWest at 10 am. I had to make a trip onto Manchester earlier in the morning and I had expected to get back to Wilmslow in ample time. Unfortunately I got stuck in a prolonged traffic jam. When it became clear that I was going to miss my appointment I wanted to phone the Branch to apologise and say I was going to be late. I didn't have a phone number for the Wilmslow Branch and so used the number from the phone book which gives the same number for all branches of the NatWest. After the usual long delays I finally asked the chap I got through to to please pass on my message that I was going to be late. But oh dear, he asked me all manner of unnecessary and irrelevant questior: my account number, the sort code of the branch which holds my account, which isn't in Wilmslow, and even my date of birth. This annoyed me very much since it took time and I was already running very late. There were only two things he needed to know. My name and the name of the chap with whom I had an appointment. Whether or not I had an account with the NatWest wasn't relevant. I might not have done and might have been perhaps only a prospective customer. Please take note of the fact that it is a great inconvenience not to publish the telephone numbers of your branches in the telephone book and also that your telephone answering people should use their human intelligence and not act like programmed automata. In the event I reached the Wilmslow Branch 45 minutes late and found that thankfully my message had reached them.

Regards yours sincerely
Dr John E.B.Ponsonby

10: OTHER PROJECTS

The Scottish Centre

In June 1998 John wrote a letter to the Bill Love Partnership(BLP), a town planning consultancy based in Taynuilt,a town a few miles east of Oban in Argyll and Bute, that he had found in the Yellow Pages. He explained that he had just taken retirement from Jodrell Bank and was keen to live on the west coast of Scotland and continue his radio astronomy by building a fully-steerable 7m dish antenna. He thought it might be good to supplement this with a Visitor Centre, since they seemed to abound in the area and might attract some funding. He recognised that there might be planning issues and he would welcome guidance from the Partnership on those. But he would welcome local advice on funding opportunities given that the Centre would have educational value and might well attract tourists.

He had no particular site in mind and no adequate funding but as John said "one has to start somewhere."

BLP responded reasonably positively and agreed to charge a fee of £600 to progress it to a stage where funding had been identified to cover further costs.

John wrote back on 1 July accepting the offer and explaining about the challenge to radio astronomy from interference and suggesting that the partnership might care to look at the CRAF website to get a handle on this. He also mentioned his two "unfinished" masers which he would plan to bring up to the Centre to complete them.

The next significant event was that John wrote a note titled

"Themes and Notions of Proposed Astronomical Centre and Space and Time Observatory" and sent it to BLP.

The main points in the note were:

- The minimum acceptable dish was 7m; better would be an 11 or 14m one
- Hydrogen masers would keep precise time
- The telescope would be used to map the 21 cm spectral line of atomic hydrogen and so the distribution of hydrogen in the spiral arms of our galaxy
- Display boards would be home-made to save money
- There would be demonstrations of the Balmer lines, electrolysis of water, hydrogen- filled balloons, heavy water, optics and an Eternity Clock
- Outdoor displays would include a scale-model of the solar system and our galaxy
- Other displays would include: GPS, radio spectrum management, organisations and something generally on the hydrogen theme
- A remote location would be essential and cars should be parked a minimum of half a mile away. Visitors would have to walk.
- There would be a shop and notice boards to display, among other things, recent astronomical events
- There would be a need for a small electronics laboratory, a darkroom, a room equipped for making one-off printed circuit boards, a workshop with machine tools and an office

In summary, perhaps, almost everything that John was interested in and dreamed of.

He went up to Scotland at the end of October to meet BLP and other potentially interested people.

In his letter thanking BLP (who had now become Bill and Sandy) he said that the prime thing for him was now the setting up of the Observatory; the Visitor Centre was just an adjunct. He even suggested that he might like to start with two telescopes: a 7m one and a 14m to be used for pulsar research, taking advantage of the two masers.

This was not music to BLP's ears. For them (13/11/1998) the Visitor Centre was the element that might attract funding. There would be very little likelihood of the Observatory alone attracting any support so John would have to resource it himself. Even planning would be affected: the combination of a house and a telescope would be unlikely to be seen as more than a residence and equipment for the pursuit of a hobby. The time they had spent on the Visitor Centre had been wasted.

Still no-one had made any estimates (even vague ones) of what it might all cost. It was an extraordinary omission given that the costs involved were probably enormous with possibly two radio telescopes and all the supporting facilities as well as the Visitor Centre. It was probably millions and far more than any likely funding body had available.

Sir Malcolm Guthrie Bt became involved at the suggestion of BLP. Briefly, in 1988 and 1989, a Formula 2 racing driver, he seems to have had a range of interests several outside the motor-racing world, including shooting and gun-making. Some of these may have given experience of working in Scotland. Whatever, he wrote a paper about John's proposal aimed at the non-scientist which was rather good. He proposed a meeting with John but there is no record of that having taken place and the matter seemed to fade away.

As for BLP on 19/2/1999 they thought it time to send John the bill for £600 plus VAT (£705) explaining that the true cost of work they had done (27 hours) was £3186. John responded by sending one third of the amount demanded (£235) saying that he thought the £600 included the VAT. BLP agreed with this and sent a corrected bill for the amended balance.

That is all there is in the file and we might judge that it all ended messily – rather as early retirement had.

There is a coda – a rather typical John one. While sailing in Scotland he met Jeremy Bray MP and some time later, when Bray was Labour shadow science spokesman, at Jodrell Bank. Bray had been a Minister of Technology in the 1960s. John had clearly approached Bray with his proposal sometime in November 1998. Bray wrote to Calum MacDonald MP, the Minister for Housing, Transport and European Affairs in the Scottish Office telling him about John's idea and suggesting that they should meet.

The reply from MacDonald was two months in coming and was "non-committal but sympathetic... as much as one can expect from a minister at this stage" (in Bray's words).

There was talk of buying a remote island but friends persuaded him that he would be lonely and it would deter any possible visitors.

John may well have decided that enough was enough. After nine months he had reached a dead end and the masers would have to stay in the garage for a bit longer.

11: ASPECTS OF JOHN

WOMEN

John attended exclusively male schools; his contacts with girls was limited to school holidays. Even then there's evidence that his contacts were closely controlled by his parents and were probably limited to his community of cousins. It was only when he went to Imperial that he developed a taste for feminine company and discovered that he was quite attractive to women. Someone who knew him from his Imperial days, Jane Pine Coffin, remembered him as a sociable (even party) person with plenty of friends of both sexes. She recalls him as a teddy bear who women wished to hug and as a good listener. There were parties and dances with all the excitement of young things eating fish and chips from newspaper. Girls sat on boys' laps; she sat on John's.

Illustration 16: Jane Pine-Coffin

Jane recalled:

> "He had a very old Morris car, which he seemed to spend a lot of time fiddling with the engine. I remember him taking me out to dinner on his 21st birthday and on to a dance at the Students' Union. I couldn't find a birthday card for him – they were all very twee or dull – but in one shop they had some small fur covered mice so I brought him that instead of a card".

Jane got into the habit of sending John a small gift in the form of a mouse on his birthday. One of these wooden mice was placed on John's grave in 2021.

When he moved to Manchester he shared a flat in Fog Lane in Didsbury[35] with Mike Quigley, a fellow research student. It was about this time that John met Ros who was training to be a physiotherapist. John and Ros seemed to get on well and they went out together for several years.

> "From 1960 to 63 I had a girl friend whom I very much wanted. You know how men are, it is very easy for a man to want a woman sexually. Generally however they are terrified of getting her pregnant. What is rare is to want to get a particular girl pregnant, to want her to be the mother of ones children. That is how I felt about this girl. But she had other ideas and I was hurt."

There are two versions of why the relationship ended: either his mother considered her unsuitable and forbade John seeing her again or Ros simply dumped him (this was John's version of events). Ros went off and married someone else. She died relatively young and

---

35 He rented flats in the Didsbury area until in bought his house in Dean Close, Wilmslow in 1973. He lived there for the rest of his life.

John, forever after, used the experience as an excuse for terminating relationships that waned or seemed to be becoming too serious for his taste.

One woman said that he charmed women and they fell in love with him but he never would do anything about it. She revealed that she had a mesmerising moment when watching John with his friends' children on his knees, longing that she could take their place.

She spoke with real affection for John.

Many of John's admirers were in his own age-group but he was liked by much younger women. And he liked that. He seemed well-aware of the attention and indeed he was rather pleased with his allure

When asked to explain the attraction the answers were: he was fun and hence good company and he was an excellent listener. In contrast, he was not inclined to talk about himself.

One friendship which was mentioned by several people was with Marylène. She arrived with John at a camp for English friends in Ann Matthias's orchard in France and, everyone was keen to tell me, shared a small tent.

Illustration 17: With Marylène in 1979

His friends thought he preferred blondes.

Later in life he befriended a blonde, divorced Russian lady who had fallen out of favour with the Soviet regime and came to Britain to work on some kind of research contract in Nottingham. Quite how they met is unclear but she wanted to marry John. In spite of the fact that John was convinced that she was only interested in money, status and acquiring British nationality, he persisted with the relationship for quite a while. No doubt he was supporting her financially

Onr friend explained that she thought John was downright rude when she first met him in 1981 shortly after she separated from her husband, a German radio-astronomer, and moved from Germany to England. In the next 30+ years he developed into the best friend she ever had. The relationship, which she insisted was never romantic, had its ups and downs and she often wondered why she found John so difficult. When he showed so much interest in moving to Scotland after retirement, she found herself thinking that she would have to go too.

It was only then that she realised just how dependent on him she had become and began to think that she should have more friends. John, perhaps sensing there was something wrong, became more attentive.

In the last few years of his life she noticed some big changes in his driving abilities: he failed to see pedestrians and became easily confused by roadside signs. When Jo diplomatically asked if there was anything wrong with his eyes, he flew into a grumpy rage.

He set off on a long continental tour in the Volvo despite her misgivings and objections and returned after a month with stories of narrow scrapes and near misses. As time passed he seemed to become more dependent on her but he often told her that he didn't like her – and she increasingly thought he meant it. There was nothing at this stage to suggest to her that John was seriously ill.

RELATIONSHIPS

One striking aspect of his relationships, mentioned by many of his friends, was that they were compartmentalised. Several of them didn't know about John's other friends until they met at the memorial event after his death. It seemed to have been a conscious plan, perhaps reflecting his social awkwardness.

I asked him once where he was spending Christmas – perhaps with friends? He replied, quite seriously, that he had no friends.

His social awkwardness made his relations rather fragile. The traumas of being a friend or even coming into contact with John were considerable.

Friends described it as a roller-coaster: great attentive warmth followed by periods of indifference, animosity and even cruelty.

To others, less than friends, he had a knack of saying something offensive, knowingly or not. It was often interpreted as arrogance and, whatever, it sometimes abruptly ended things. People went away bearing a grudge or writing it off as just John being John.

There were endless accounts of John's ability to fall out with people: looking up from a conversation round a table in a restaurant to see John having a heated argument with a waiter, a *maitre de* or a hapless member of the public. When asked what it was all about John would generally say something like "The man's a fool" perhaps adding some comment on the person's qualifications (or lack of them).

John's friend Mac told me about the incident of the glider inspection.

Mac had a glider which needed to have its annual check by an independent inspector. He invited John along.

The inspection proceeded but John started fiddling about with parts of the glider and pointing out potential faults to the inspector even sometimes after the man had OK'd the part. It seemed to go on, to Mac's frustration, for a long time.

When challenged afterwards John said that he thought he was being helpful – making the inspection more thorough. Mac didn't see it quite that way.

John joined the committee of a sailing club, invited because he seemed to have new ideas. While some of the members accepted his rather abrasive manner (putting it down to his being a "bit of a one off" and a scientist) a few found him simply arrogant and rude.

One person took it particularly badly when John made, in his usual dismissive way, a critical remark about a book the member had written about the club's history. He swore never to speak to John again. He later relented when, by some error, they had to sit together at a dinner and John proved an entertaining companion.

Another involved the Secretary of the committee and started, in the way these things often do, with a clumsy email from John suggesting (as in demanding) some kind of change in procedures. The Secretary was deeply affronted and refused to send John ("that man") any committee papers in future.

As a result they had to be sent to John surreptitiously. John turned up at a few meetings after that, apparently unaware of the distress he had caused the Secretary but it was noticeable that none of his fairly frequent contributions actually made it to the Minutes.

Perhaps sensing that all was not well, John stood down at the first opportunity.

When he met a young Turkish student who was doing a PhD in accounting (devising a new pension system for the Turkish government) he was eye-wateringly rude, persistently asking the young man how accountancy could possibly be a worthwhile and proper subject of study.

He approached a close friend to confide that on some occasions, when he turned up somewhere unannounced and expecting lunch, he didn't feel particularly welcome. The friend assured him that he was always welcome but that the targets would probably appreciate some warning of his approach.

John seemed to appreciate this but he stuck with the old system.

On another occasion John lent his mother's accordion (being French she had one) to a friend. When he came to want it back, he tortured himself for a long time about how to ask for it. He eventually discussed it with another friend. Seeing how distraught John was, this friend offered to sort it out.

When the accordion was returned John asked him what diplomatic magic he had employed and the friend replied:

"I said John would like his accordion back".

There seems little doubt that John was regarded by most people who knew him well as neuro-divergent: his brain was wired atypically. He would have said "Am I?" and asked for a definition but, probably, he would have liked that.

Several people mentioned autism and were convinced that John was on the spectrum. As someone observed:

> ",...because you want to be in your world, but the real world doesn't live up to your expectations. And again, this may be this feeling of being slightly abandoned as a child as well, and not getting the love that he may have wanted."
>
> "You know, if you don't have an example of love around you, and what a proper, if there's such a thing, relationship it was probably difficult for him to sustain a relationship".
>
> "That's why he loved being with Marina, I think he may have even slept on the floor or on a sofa bed and there were hundreds of them everywhere but he just liked that feeling of being in a family."

CHILDREN

John had an affinity with children. He seemed comfortable with them particularly once they seemed ready to take an interest in science. He would string resonant pendulums across rooms and garages to entertain them or have some gadget or toy about his person to intrigue them.

Illustration 18: Balloons in liquid nitrogen

He often turned up with a thermos flask of liquid nitrogen and had great fun immersing un-inflated balloons in it and then pulling them out. They inflated as if by magic. Another favourite he carried about with him was his Levitron.

This little device is composed of a base magnet and a spinning-top which is also magnetic. Using the principle that like magnetic poles repel, the top can be spun and it will levitate over the base magnet – with no visible means of support – for a minute or so before slowing down and falling off.

Irma Kennaway reflected:

> "He loved driving his green enormous Volvo to Como. He would bring all his scientific instruments, put them on my terrace, hang all sorts of dangly things out windows, and any child walking past because you can see the terrace from the lower street would see an enormous telescope on our terrace and get absolutely fascinated. And he would always bring

incredible scientific games and was incredibly concerned about children and I followed a young Moroccan boy since he was 13 and JB came to the children's home with me and met this Moroccan boy and he brought one of these games where something spins in the air and stays up. Of course this boy was fascinated and John always wanted to amuse everybody. And he had a little tin boat that if you lit a candle on it, it would then whizz around the bath.

And I felt there was a big child in him. I mean, I'm an artist, so there's a big child in me. He just, this playfulness, this wanting to interact with children. This, you know, when I went to the children's home wanting to bond with this young Moroccan his eyes would light up. So maybe there was that, you know, suppressed sadness and anger.And from what I've heard, he's left all his money to very young people to give them a hand."[36]

Children seemed to like him because he was not your usual adult. Friends who took their children on skiing holiday with John recall mornings where John would be sleeping under a heap of blankets and warm coats on the floor oblivious to the circle of children who gathered around waiting for him to wake up and come skiing with them. Eventually, the heap would start to heave and a sleepy John would half emerge scratching the top of his head a la Stan Laurel. The shrill cry would come "Are you coming skiing with us John?". The invariable bleary response would come from John "Am I?". His strategy on these occasions was to answer a question with another one.

DEPRESSION

Several people mentioned John's depressive nature – sometimes referring to Churchill's "Black Dog".

---

36 From Irma transcript – slightly amended

He certainly had periods of moodiness and depression when he would retreat to his house for a few days but he usually returned, irascible for a while perhaps, to his normal world.

He seems to have been going through some kind of therapy from time to time. It was possibly for depression but there were several other possibilities.

> "Thinking about the childhood he had, how on earth that he managed to hold down the job for so many years, how on earth that he managed to have so many friends, and be so actually together that he could, you know, get out of bed in the morning and drive himself to Italy and, and not be complete depressive or, you know, let himself go all together. So I always thought, well, how do you do it?
>
> I used to call him a lot and say, how are you doing? And he would say, all right, and then always go back to his boat. It was his boat that was keeping him ticking over."

ANGER

John was liable to bouts of anger if things did not go as he thought they should. One incident which was mentioned by several people – even by people who were not actually present – was the scattering of the chairs in the Jodrell Bank canteen. The details are difficult to come by: memories fade and legends grow.

The bare facts are simple enough. An apparently gentlemanly conversation was going on at lunchtime when John abruptly and furiously stood up, uttered some expletives suggesting he had been misrepresented and strode off, scattering the said chairs before him. Another version has John not just scattering the chairs but smashing them.

The other people in the canteen at the time no doubt looked up, saw it was John behaving badly and seamlessly returned to whatever radio-astronomers talk about when they're having a break — radio astronomy probably.

The cause of the outburst is, to this day, a mystery. Some minor difference in the interpretation of the maximum entropy principle, a chance remark about masers, a complaint about parking perhaps. A taunt about United's performance at the weekend never.

John must have been one of the few people who have flown into a rage over empty jam-jars. In his final days someone decided to throw away a fraction of the hoard of these which added a dusty sparkle to so many rooms in the house. When he found out, John was beside himself with outrage: his jealously-guarded hoarding rights had been infringed without his permission.

The trigger for these disproportionate outbursts could be tiny. A neighbour, who regularly did his shopping when John wasn't able to, recalls returning from the supermarket with a large shop. When John discovered that one item was wrong (it may have been Cheddar cheese rather than Cheshire) he flew into an abusive rage, totally ignoring the fact of the neighbour's months of kindness.

Ann M had a similar experience with salt. John was in the Christie and decided that he wanted a salt cellar. Ann was doing his shopping and was asked to buy one. It was that time between Christmas and New Year when things like salt cellars do not take centre stage on the shelves and Ann had great difficulty in finding one. After scouring the shops of Didsbury she eventually found something that seemed to fill the bill. It was a salt mill filled with crystals of sea-salt. Feeling rather pleased (sea salt is, after all, claimed to be more flavoursome than the regular kind), she presented it to John in his sick bed.

John flew into an apoplectic rage claiming it was the wrong sort of salt. He rang Flossie in Ipswich to complain.

Later he was to admit that he liked it.

He could be somewhat confrontational unexpectedly. A plumber who claimed to be a plumbing engineer was asked to explain what degrees he had in the subject. As the plumber paused to consider this novel question, John told him that *he* had three in his field.

It could provoke responses. A friend said this:

> I've known us walking into a shop, and I might be looking at something, and the next minute he and the shop owner had gone about each other a hundred times. He could immediately get somebody irate.
>
> I kind of feel he was this man with a golden heart, and living in his own world, and I'm not really in tune with the mortal world, you know, I think he was in his own world. That kind of made you feel you wanted to protect him, Yes, he was very vulnerable.
>
> As if he was trying to protect something of his and that probably came across with the angry and rugged side that there was something inside that he didn't want everybody else to see.

PERFECTIONIST

John was a perfectionist: if something was to be done, it had to be done his way. This had several effects.

Things just didn't get finished. The maser saga went on so long and without real issue partly because it turned out to be much more difficult than anyone expected but also because John seemed unable to compromise. He refused to listen to other people who might have helped and did it his way. People eventually walked away and left him to it.

A side-effect of his attitude was that he alienated anyone assigned to him.

One story I was told was that he had asked a technician to do some wiring of a device. The man worked diligently for several days to complete this fairly complex task and presented it to John. John came in the following day with a pair of garden shears and hacked the whole thing apart saying that it was untidy. It was a dreadful act in cold blood but news of it soon spread around the lab. It was John being John to many of his colleagues. To some (mainly technicians no doubt) it seemed like a spiteful, humiliating act. Small wonder that people didn't like working for him.

It went beyond work. He worked on his boat for nearly 30 years on the land making large and small improvements; tinkering and varnishing. He never quite finished anything and she was never ready to go into the water again. Paradoxically, the boat was in a mess when he gave up on her.

Peter told me:

> I discovered that one of the drains at his house was blocked outside, which wasn't great. So I thought, I'd speak to Michael [a neighbour] and not mention it to John because I knew if John got involved things would have to be done his way, or not at all.
>
> So Michael and I secretly unblocked the drain. I don't think John was ever aware that it happened. "

Ann M found that John was very ready to fix things in her house but perfection takes time and the smallest job seemed to involve several visits – and several free dinners

It was suggested to me that perfection for John meant perfection in process rather than in outcomes. It explains a lot.

HOARDING

Jo described her effort to tidy up his house in the late 1980s:

I gave up my life and devoted myself to John's house. He wanted it doing, basically. And he joined in. I remember him on a hot day, standing up a stepladder just with just shorts on and papering ceilings. Really? He really enjoyed it. We went to his back bedroom. Oh, and you could hardly see. It was terrible. It was black cobwebs and the paper was hanging in loops. And the curtains that were there had been silken. They were rotten. There was a built-in wardrobe that a previous owner had made. And I opened it. Things like duffel coats had fallen down. And they were all covered in green. It was terrible, terrible. Absolutely awful. And we threw 18 bags out that day just for that room. And then finally stripped the paper off and re-papered it and made it into a wonderful computer room for him.

I worked my way through every room. And did enormous work. For two years. I just devoted myself to John's House. And we just had the kitchen left. But then his mother died in 1989 and he started visiting Thame and coming back with carloads full of stuff. Now, you couldn't open or close the door for bookshelves. So I put them on another wall. And rearranged the whole thing.

We made it lovely. And there were two easy chairs. And a little coffee table in between. And I made everything so nice. But then he was bringing all these occasional tables. He got one occasional table under the window with a couple of chairs. If he wanted to sit at the window. But the room just got full of stuff.

He brought his mother's bedding on one trip. I don't wish to malign the old lady but she didn't know that you could wash blankets.

And in the end I just couldn't do any more. I just had to retire. Because it was like swimming in treacle. Fighting John. Trying to explain why he didn't need those things

I never thought he was grateful for anything I did.

After this things went downhill and John kept most things: tiny bits of wire that might be useful, jam jars, books and journals, paintings of, generally, his Ponsonby ancestors and broken crockery that he had mended. I found a full-sized rubber dinghy under the newspapers that carpeted one room.

The garage, which was connected to the house central-heating system to preserve the two masers, contained the masers (which seemed to be permanently plugged-in to the mains supply), a numerically-controlled metal-working lathe, miscellaneous tools, a couple of masts for dinghies and shelves of boxes containing, mainly, pre-used wood screws. The floor was littered with pipes, rods and general impedimenta so it was necessary to leap between clear spaces to get anywhere.

One curiosity was a small bottle filled with heavy water which was somehow associated with the masers.

Round the back of the garage there were the rusting remains of a framework for a ferroconcrete boat that John had intended to build in the 1970s.

Luckily Johns was (heavy water apart) a dry hoarder.

Peter Twyman , Rosalind's brother, came to look after John for a while after he was discharged from hospital. Peter told me:

> I was having some difficulty because there was nowhere in the house to sit down except at John's desk. And so if I wanted to sit down, I'd sit down on the bottom of the stairs. There were books and all sorts of things on the stairs all the way up, but there was enough space for you to walk up and down or sit down indeed.
>
> In the kitchen there were just no clear surfaces left, and there were innumerable empty glass bottles everywhere.
>
> So I spoke to him before he came home and said, do you mind if I get rid of all these empty glass bottles because the kitchen's unusable at the moment? So he said, that's OK. So I think I got rid of two wheelie bins full of empty glass bottles

and managed to clear two spaces, one next to the cooker -- which was less than two feet square – and one next to the draining board on the other side of the kitchen – which was of a similar size, maybe slightly larger. But there was nowhere in the house we could sit down and have a meal together.

There was just no way he would throw things away but I did actually manage to organize things. I discovered a high stool, sort of buried in one of the rooms and I could sit on it in the kitchen. To begin with, I would take him food and serve him where he was sitting at his desk. But he didn't like that. He wanted us to eat together, which was understandable. But he would then, when a meal was ready, come through to the kitchen, and he'd stand, and eat it standing while I sat on my high stool. It was only in the last couple or three weeks, I suppose, that he was prepared to sit on the stool while I stood. "

When asked why John hoarded there were various answers but most of them were essentially the same. John led a rootless life for most of his childhood and teens: boarding in a variety of establishments, his parents absent and permanently on the move. It meant that he generally had rather few personal possessions around him.

When he finally settled down in his own house he made up for it by never throwing things away.

RECORDING

It seems somehow related to his hoarding behaviour that he kept records of mundane matters. If he travelled for any distance in his trusty Volvo he made a note of the stages

"3:17 Hilton Park Services 285 miles"

When he went abroad he made detailed itineraries specifying

hotels he was to stay in and people he was going to meet (usually with their telephone numbers). They were usually typed. He kept all the receipts. All this in neatly-labelled files, safely hoarded.

GENEROSITY

John was a generous person in the financial sense of the word. He supported numerous people, family and friends, with significant loans (which were either repaid or not) and outright gifts. They included payments to people who were encountering temporary hardship, support for someone's education and a loan to someone else to start a business.

In particular, he took his role as god-parent very seriously and was usually ready to be very generous when the need arose.

According to neighbours John was vulnerable to doorstep sales-people with a sad or plausible tale to tell. Itinerant tarmacers offering to sort out his drive or others to pressure-wash his paths – familiar with rejection – were often welcomed. If they had some appreciation of the science of what they did they usually scored. So lucky heather sellers may have had little success.

Of course, he was a relatively wealthy man so he could risk a bit of altruism. His own needs were meagre (unless it came to a numerically-controlled lathe to work on his beloved masers after retirement). His general indulgences went hardly further than a take-away kebab on a Friday night.

He was not necessarily generous when it came to the smaller rituals of social life. An instance was that in his practise of dropping in on his friends for a meal uninvited, he never brought a bottle of wine, explaining that he seldom touched the stuff. Yet he seemed to be readily pressed to drink several glasses.

Ann M became so irritated by this that she made a list of acceptable, reasonably-priced wines available from Victoria Wines (a shop that was next door to where he lived at the time) and gave it to him. Alas, she couldn't remember if it worked.

When his accountant told him that he was actually a rather wealthy man and didn't really need to live as frugally as a student he replied with a coy "Am I?".

When he learned that the brother of a close friend was in financial trouble while setting up a business – his bank wouldn't lend him any more – John rang him and asked if he could help. When £80,000 was mentioned John agreed to lend it to him. Subsequently he lent him another similar sum. It was later paid back but there didn't seem to be any real pressure to do so.

He supported many other people, family and others, at home and in France and Chile with loans and straightforward gifts.

When he died his estate was just shy of £2million. There were 37 individual beneficiaries (cousins, godchildren and the grandchildren of friends) in his will and three charities. The three charities: Orbis, Greenpeace and the Noel Buxton Trust benefited to the tune of about £500,000 each.

Quite how he managed to acquire so much wealth, much of it squirreled away in obscure – sometimes foreign – bank accounts, is something of a mystery.

LEGACY

Professor Simon Garrington, Associate Director at Jodrell Bank, wrote this appreciation of John's work:

> John was one of the technical pioneers at Jodrell Bank from the early 1960s onwards. He built the 'Doppler Ephemeris Machine' for planetary radar experiments and carried out one of the first (1961) radar determinations of the distance to Venus and hence the Astronomical Unit. He developed early techniques of synthetic aperture radar for imaging the Moon, and also made important early contributions to techniques of aperture synthesis including Maximum Entropy. He spent

> many years designing and building a sophisticated Hydrogen Maser frequency standard at Jodrell Bank, which was used experimentally. John had a fierce intellect which he applied to a wide range of challenging problems. His hand-written one-page notes were brilliantly illuminating. He was a force to be reckoned with and an unforgettable character, and one of the true pioneers of radio and radar astronomy at Jodrell Bank.

You would expect a generous formal tribute in an obituary but there is little to quarrel with in this assessment. It is though only part of the story. The underlying one is of great scientific and engineering achievement in spite of many features of his own makeup, some of which he struggled with and some of which he probably never recognised. His arrogant attitude and unwillingness to listen to and act on the advice of others made him difficult to work with. After a dose of "John knows best" many people shrugged and walked away. Only people who recognised his exceptional technical talents and his commitment – and saw the generosity and humanity underneath – stayed.

In career terms he was lucky to arrive at Jodrell Bank at the inception of the radar work and slightly unlucky to find that after 10 years Jodrell decided not to go on with it. It was though probably his most cited work.

The next significant project was the masers which could have taken him straight into the multi-telescope field. He may have recognised it as a challenge involving many disciplines beyond the usual realms of electronics but he surely did not appreciate just how big the challenge was particularly since the maser was required on such a short timescale.

He may well have been expecting a supporting team which he would lead. That seems not to have materialised for whatever reason and he worked with little technical support throughout.

Important deadlines came and went as continual problems arose – no doubt partly due to John's perfectionist tendencies and his inability to accept advice – and eventually Jodrell had to buy a commercial device for the planned project.

John was then allowed to continue with development work, off and on, on his two masers for the rest of his career. He was, even then, expecting to continue into his retirement.

There is some evidence that they never did work properly anyway-so Garrington chose his "was used experimentally" rather carefully.

It is all to easy to blame John for the fiasco but much of the fault must surely lay with the management who chose John to do the job, failed to monitor progress and didn't do anything when it was clear that there were enormous problems.

John was a victim rather than a perpetrator. The stress he went through, given his high expectations of himself and his naturally depressive nature, must have scarred him for the rest of his career.

Nonetheless he carried on with his usual enthusiasm doing many other innovative things until disappointment with lack of advancement and disillusion with the direction Jodrell was taking crushed his dreams with an offer of early retirement.

He left with no great body of work to his name as a legacy because his range of interests was so wide and electronic engineering made so many strides so quickly that his contributions were quickly overtaken.

His stoic defence of frequencies for radio astronomy, with his colleague Jim Cohen, in his last few years at Jodrell, may be his lasting gift to the world. We shall see.

Most people seemed to agree that he was not a team player at work but a loner who thought he knew best – which he frequently no doubt did.

Some of his friends pointed out his difficulties with understanding personal relationships as the root of his problems. It could manifest itself in different ways: his fits of anger if he felt his views were being ignored or contradicted, his careless rudeness, his arrogant manner, and his insistence that everyone should know that he was the cleverest person in the room. But his charm and sociability with people he knew, his willingness to help people in trouble, his generosity showed another side of him.

Some people said he was clearly on the autism spectrum; others had never thought of that possibility. Some wondered whether there was no lack of understanding but that his essential arrogance made him careless about the impact of his actions other peoples' feelings. It seemed somehow linked to his conscious compartmentalisation of his friendships: people very close to him met one another for the first time at his memorial event.

One curious and unexpected feature of his life was his many close (but often from afar) relationships with women. I counted half a dozen from the correspondence. They seemed to generally (but not always) start when the woman was having difficulties of some kind. John would listen and support her (sometimes financially). She would frequently declare her love and seek an even closer relationship. Usually John would start looking for a way out shortly after this.

Sometimes the exit strategy was to ignore her; sometimes things ended in cruel remarks from John. He often used the excuse of Ros (she of the early days at Jodrell who had dumped him) as perfection, the love of his life and the only women he ever wanted to have his children.

Whether any of these affairs ever developed any sexual aspects nobody mentioned and I didn't like to ask. It was none of my business and just another part of the enigma that was John.

Most people who knew him well commented on his childhood and its likely effect on him. His father was 47 when John was born, his mother 35. He was quite probably conceived as his parents floated down a steamy Congo river on their honeymoon. It seems that there wasn't much steamy after that. According to John they hardly touched one another.

The expressions of affection seem to have been limited to questions about John's school performance: John's letters home from his various boarding schools concentrated on his performance in tests and games.

The few people who remembered his parents recalled them as trapped in Edwardian, colonial ways of restraint, suspicion and snobbery. John felt little connection with his father and expressed doubts about his love for his mother.

What he must have felt though was the great weight of expectation that he should succeed in life – he came from a high-flying family of

Ponsonbys after all. He probably inherited it and expected a lot of himself. His early travels in Africa and to South America made him quite different from his classmates – more urbane. And of course he was half French (or was it German?)

At his early schools he seems to have been troublesome and friendless. At Marlborough, after a shaky start, he at last settled down and fitted in. In the way of great public schools his expectation was transformed into a sense of entitlement and a growing self-confidence.

The fact that he had often spent his holidays at Escot would have helped. When he was there he had a nanny and Aunt Felicity. She seemed a lot more fun than his parents

So his childhood was a fragmented one: shuffled around nurseries, kindergartens and prep schools, his parents often absent. Miserable, not fitting in and probably bullied.

As one person put it:

> "considering what he went through in his childhood, he was surprisingly normal."

If this thought had ever been put to John he would have said, in his usual rather distracted manner, "Am I?"

He would, after an irritated pause, not liking the sound of that, have asked "What exactly do you mean by normal?"

APPENDIX 1: FAMILY TREES

1: The Ponsonby Ancestors

2 & 3: The Descendants of Stewart Ponsonby and Mary Buxton

4: The Franco-German connection

TREE1

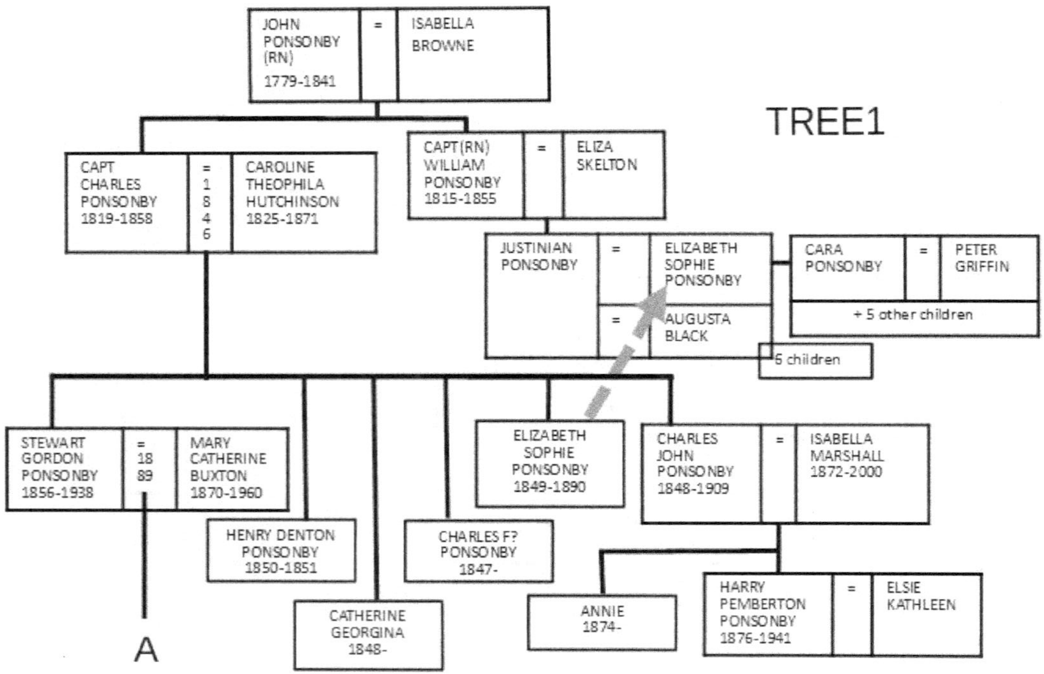
JOHN PONSONBY (RN) 1779-1841
=
ISABELLA BROWNE
CAPT CHARLES PONSONBY 1819-1858
= 1846
CAROLINE THEOPHILA HUTCHINSON 1825-1871
CAPT (RN) WILLIAM PONSONBY 1815-1855
=
ELIZA SKELTON
JUSTINIAN PONSONBY
=
ELIZABETH SOPHIE PONSONBY
=
AUGUSTA BLACK
6 children
CARA PONSONBY
=
PETER GRIFFIN
+ 5 other children
STEWART GORDON PONSONBY 1856-1938
= 1889
MARY CATHERINE BUXTON 1870-1960
A
HENRY DENTON PONSONBY 1850-1851
CATHERINE GEORGINA 1848-
CHARLES F? PONSONBY 1847-
ELIZABETH SOPHIE PONSONBY 1849-1890
CHARLES JOHN PONSONBY 1848-1909
=
ISABELLA MARSHALL 1872-2000
ANNIE 1874-
HARRY PEMBERTON PONSONBY 1876-1941
=
ELSIE KATHLEEN

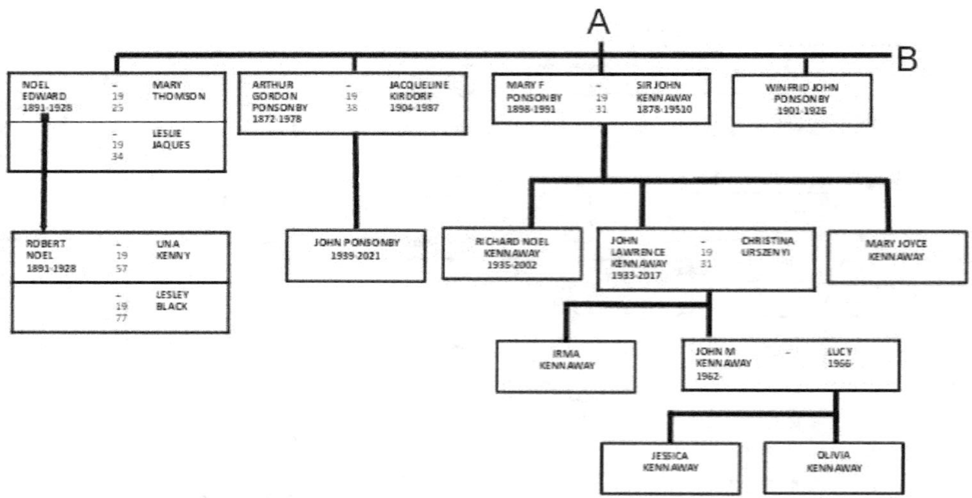
A
B
NOEL EDWARD 1891-1928 - 19 25 MARY THOMSON
- 19 34 LESLIE JAQUES
ARTHUR GORDON PONSONBY 1872-1978 - 19 38 JACQUELINE KIRDORF 1904-1987
MARY F PONSONBY 1898-1991 - 19 31 SIR JOHN KENNAWAY 1878-19510
WINFRID JOHN PONSONBY 1901-1926
ROBERT NOEL 1891-1928 - 19 57 UNA KENNY
- 19 77 LESLEY BLACK
JOHN PONSONBY 1939-2021
RICHARD NOEL KENNAWAY 1935-2002
JOHN LAWRENCE KENNAWAY 1933-2017 - 19 31 CHRISTINA URSZENYI
MARY JOYCE KENNAWAY
IRMA KENNAWAY
JOHN M KENNAWAY 1962- - LUCY 1966-
JESSICA KENNAWAY
OLIVIA KENNAWAY

TREE2

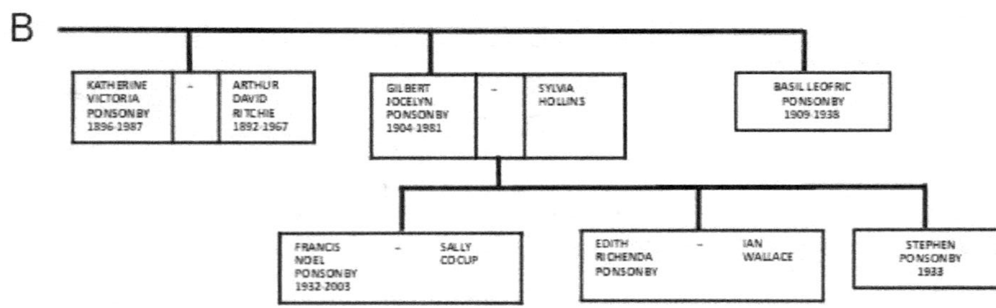
B
KATHERINE VICTORIA PONSONBY 1896-1987
-
ARTHUR DAVID RITCHIE 1892-1967
GILBERT JOCELYN PONSONBY 1904-1981
-
SYLVIA HOLLINS
BASIL LEOFRIC PONSONBY 1909-1938
FRANCIS NOEL PONSONBY 1932-2003
-
SALLY COCUP
EDITH RICHENDA PONSONBY
-
IAN WALLACE
STEPHEN PONSONBY 1933

TREE3

TREE4

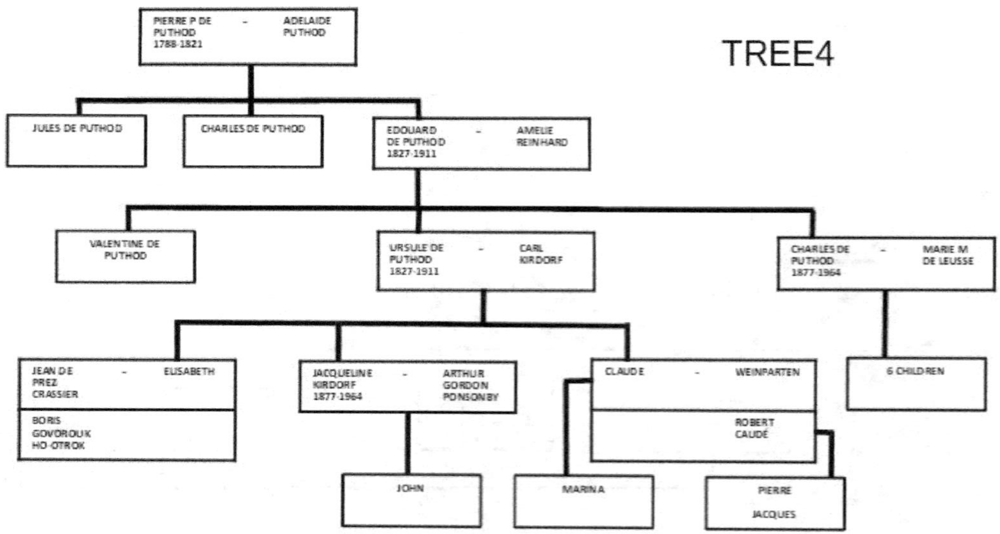
PIERRE P DE PUTHOD 1788-1821 - ADELAIDE PUTHOD
JULES DE PUTHOD
CHARLES DE PUTHOD
EDOUARD DE PUTHOD 1827-1911 - AMELIE REINHARD
VALENTINE DE PUTHOD
URSULE DE PUTHOD 1827-1911 - CARL KIRDORF
CHARLES DE PUTHOD 1877-1964 - MARIE M DE LEUSSE
JEAN DE PREZ CRASSIER - ELISABETH
BORIS GOVOROUK HO-OTROK
JACQUELINE KIRDORF 1877-1964 - ARTHUR GORDON PONSONBY
CLAUDE - WEINPARTEN
ROBERT CAUDÉ
6 CHILDREN
JOHN
MARINA
PIERRE
JACQUES

APPENDIX 2: MASERS

One of John's major and continuing preoccupation was with masers and their role in precise timing, which was key to success in long baseline interferometry.

The simplest model of an atom is that of a nucleus composed of protons and neutrons surrounded by electrons in a variety of shells or orbits around it. These shells correspond to higher energies as they get larger and they are quantised. In other words the energies have discrete values and there is a kind of ladder.

If we excite an atom with, for example – if it's a gas – an electrical discharge, we will either knock an electron into a higher shell or remove it entirely from an atom. If the electron is knocked into a higher shell then the remaining electrons will rearrange themselves fill the hole. The lost energy will be emitted as one or maybe more photons. Since the energy levels are quantised (steps in the ladder) the photon energies will be discrete and characteristic of the energy transition that the atom has undergone.

One of the early results that led to quantum theory was the observation that there is a simple relation between the energy of photons and their frequency:

$$E=h\nu$$

where E is the photon's energy, ν is its frequency and h is Planck's constant, named after the man who made the connection.

The frequency of the photons may be anywhere between the infra-red and well into the x-ray spectrum depending on the element concerned and which energy transitions are involved. The spectrum of an element is so characteristic that it can often be used to identify elements in distant stars and galaxies.

In fact when you look very closely the discrete energy levels have a fine structure and are made up of several distinct levels. If you look even closer it emerges that there is a hyperfine structure. The hyperfine structure arises because electrons and protons have a property called spin, which can by convention either be up or down. Being electrically charged they generate a magnetic field – up or down. If the magnetic fields of the nucleus and electron are both up or both down the system has a slightly higher energy than if they point in opposite directions. When ever an electron flips from the higher energy state to the lower one, it emits a photon and because the energy gap is so small this is of much lower frequency. For hydrogen and the transition of most interest to us, the frequency lies in the microwave region. Its wavelength is about 21 cm and its frequency is 1.420405751768 GHz – a candidate surely for one of the best known quantities in science.

Maser is the acronym standing for Microwave Amplification by the Stimulated Emission of Radiation. So we now need to look at what stimulated emission means and that will help explain what amplification has to do with it.

When an electron is in the higher energy state it can spontaneously and randomly drop to the lower energy state and emit a 1.42 GHz photon – helpfully this is called spontaneous emission. However if there are already plenty of 1.42 GHz photons around it is more likely to switch: the more photons there are, the more likely it is to switch and itself emit a 1.42 GHz photon. The process is called stimulated emission and was first proposed theoretically in 1917 by no other than Albert Einstein.

The possibility of using it for amplification of 1.42 GHz photons are clear but it took several decades to find a practical application.

Before we look at that we should look at the concept of population inversion.

Stimulated emission competes with absorption. If there are no atoms in the higher energy state then there will be no candidates for stimulated emission. At the energy gap we are talking about there will always be some because of thermal effects but what we need for amplification is for there to be more in the upper state than in the lower one: a population inversion. In the hydrogen maser this is achieved by first of all creating an electrical discharge in the hydrogen gas to split it into single atoms and produce a lot of these in the high energy state. These then escape from an aperture as a beam which then passes through a magnetic state selector. This separates out and discards the ones in the lower energy state and admits those in the higher energy state into a storage bulb. The storage bulb is lined with PTFE since this allows the high energy state to be maintained through collisions with the walls and supports the inversion. So within the bulb we have a population inversion and thus the possibility of maser action. The storage bulb is placed in a resonant cavity tuned to a frequency of 1.42 Ghz and so the stimulated emission and maser action begin and are sustained for as long as the hydrogen lasts. Stimulated radiation has the remarkable property that all the photons it produces are not only of the same very stable frequency(1.420405751768 GHz in our case) but they are all in phase and so form the basis for an extremely accurate timing system.[37] Hence the interest in measuring it so accurately

37 The amplification possibilities were put to use as very low noise preamplifiers for radio telescopes until the Field Effect Transistor came along.

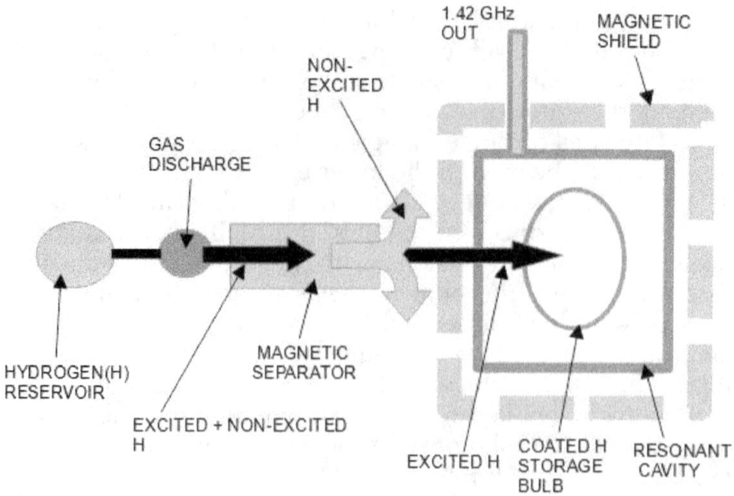

The 1.42 GHz radiation is draw off by a waveguide and processed to give a time signal that is accurate and stable over long periods. A stability of 1E-15 over a period of a day is frequently achieved, adequate for astronomical applications such as long baseline interferometry. Over longer periods the frequency tends to drift more.

We have described a passive maser because it is the usual type for precision timing applications. There is a version (an active maser) based on a quartz crystal which is cheaper to build as well as being smaller and lighter (and so suited to space applications).

APPENDIX 3: **AUTOCORRELATOR**

There are two ways that you can obtain frequency spectra in radio astronomy. The first is to use filters that pass contiguous narrow frequency windows spread around the range of interest and the second is to use an autocorrelator.

If there is a time-varying signal x(t) then the autocorrelation coefficient R(τ) is the expectation value(the average) of x(t).x(t-τ) taken over the duration of the signal where τ is the time lag . The significance of it is that it is related to the power spectral density S(f) ; S(f) is the Fourier Transform of R(τ). S(f) is pretty much the frequency spectrum we are looking for.

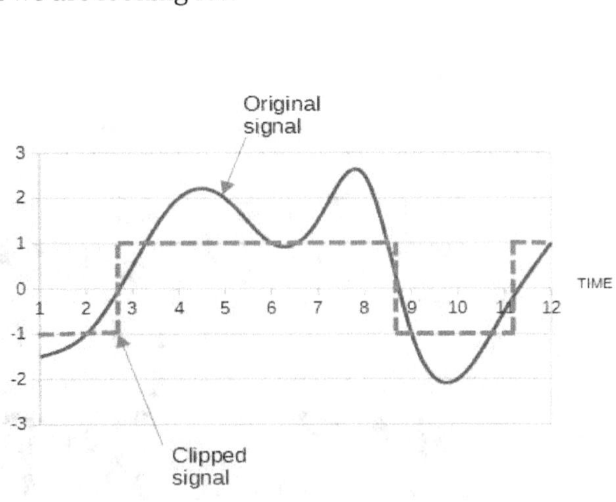

There is one remarkable factor that makes all this much easier to do digitally. It is that it works if the function is replaced by by a new much simpler, function: one that is 1 when the function has a positive value and -1 when it is negative. It is illustrated in the figure opposite.

The multiplication becomes a rule:

> If the current value of the clipped function is the same as the new function in the time sequence add 1; if they are different add

The challenge is now a binary one and so much easier to handle electronically. It was a one-bit autocorrelator.

This principle was first used by Weinreb in 1963 to make a spectrometer but it was rather limited in that it had only 21 channels and was limited in frequency range. John's would have 256 channels and would have an adjustable frequency.

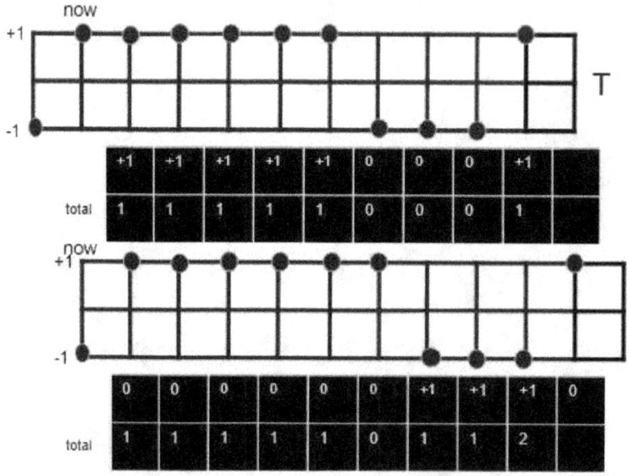

Let's imagine we have a clock that ticks once a second. John built electronics that could store this clipped signal into its current value and that at all of the previous 256 ticks of the clock (figure on the previou page) Each of the intervals had a counter associated with it. Looking at each of the 256 previous signals, if its value is the same as the current value, the corresponding counter is increased by 1; if it's not then the counter remains the same. The top figure below shows the result of the first pass after all the counters are initialised at zero.The clock now ticks and so we have a new current signal. The old current signal becomes the first previous one and the past signals moves one place to the right as shown in the lower half of the figure. If we then apply the rule we get the results shown in the figure and the total in the counters become the figures labelled as "total".

If we repeat this many times, we will build up a representation of R(τ), the autocorrelation coefficient for each second. Of course, we would need to set the clock to tick much faster t than once a second to match the frequencies in the signal: John's clock could be set to run at between 0.1 and 10 MHz. It took only a few minutes to

We have used the word "representation" because we are using just the clipped signal and not the real one. Luckily the correction factor that needed to be applied had been worked out in 1943 by a man called Van Vleck[38].

Since the data was fed into an Argos computer it was not too difficult to make this adjustment and to perform the Fourier transform necessary to produce the frequency spectrum.

A few important practical features have been left out such as the filtering system that made it possible to work in different frequency band, giving it more flexibility than previous versions.

It was tested as a spectrometer using the Jodrell Mark II telescope in several real situations and gave good results for the time, as reported in the paper.[39]

The provision of a dual input made it fairly easy, with two inputs,

---

38 There's a good chance that John didn't know about it until it was re-published in 1968

39 The spectrometer was used in studies of a pulsar

to use it as a correlator for interferometry.

APPENDIX 4: JOHN'S PAPERS

A new determination of the solar parallax by means of radar echoes.
J.H. Thomson, J.E.B. Ponsonby, G.N. Taylor, R.S.Roger.
Nature 190,519,1961

Radar observations of Venus and a determination of the Astronomical Unit.
J.E.B. Ponsonby, J.H. Thomson K.S. Imrie,
Mon. Not. Roy. Ast. Soc., 128,1,1962

Rotation rate of Venus measured by radar observations.
J.E.B. Ponsonby, J.H. Thomson,»K.S. Imrie
Nature 204,63,1964

Active all-pass filter using a differential operational amplifier.
J.E.B. Ponsonby. Electron Lett., 2, pp 134-35, April 1966.

A simple Image Forming Technique suitable for Radio Astronomy.
M.M. Komesaroff, J.E.B. Ponsonby.
Proc. A.S.A.,1,(2), 64-66, Nov 1967

Two-dimensional aperture synthesis in lunar radar astronomy.
J.H. Thomson, J.E.B. Ponsonby.

Proc. Roy.Soc. A. 303,477,1968

Measurement of the distance of Pulsar CP0328.
G. de Jager, A.G.Lyne, L.Pointon, J.E.B. Ponsonby
Nature 220,128,1968

The Jodrell Bank radio frequency digital autocorrelation spectrometer.
R.D. Davies, J.E.B. Ponsonby, L. Pointon, G. de Jager.
Nature 222,933,1969

Radar images of the Moon at 75 and 185 cm wavelengths.
J.E.B. Ponsonby, I. Morison, A.R. Birks, J.K. Landon.
The Moon, 5,286,1972

A sampling phase rotator for removal of Doppler shift in a lunar radar.
J.E.B. Ponsonby, I. Morison.
Electron Lett., 9, p409,1973

An entropy measure for partially polarised radiation and its application to estimating radio sky polarisation distributions from incomplete aperture synthesis data by the maximum entropy method.
J.E.B. Ponsonby
Mon.Not.Roy.Ast.Soc., 163,369,1973

The nature of the invisible distribution generated by the Maximum Entropy method.
J.E.B. Ponsonby
Proc. S.P.S.E. Conference on Image Analysis and Evaluation, ed. Rodney Shaw, Toronto. July 1976

Quadrature hybrid made from transmission line of single characteristic impedance.
J.E.B. Ponsonby

Electron. Lett., 20,18, pp725-726, Aug.1984

Metre-Wavelength VLBI experiment at Ooty
S.Ananthakrishnan et al
Indian Journal of Radio and Space Physics. 14, pp17-20, Feb 1985

A simple Loran-C timing receiver for supporting VLBI.
J.E.B. Ponsonby.
Proceedings of the second European Frequency and Time Forum, Neuchatel, pp341-349, March 1988

Shapes of storage bulbs of hydrogen masers for maximum filling factor.
J.E.B. Ponsonby
Frequency Standards and Metrology, Proceedings of the 4th Symposium, ed. A. de Marchi, Ancona, Sept. 1988

VLBI Observations of the Nuclei of a mixed sample of bright galaxies and quasars at 327 MHZ.
S. Ananthakrishnan et al.
Mon.Not.Roy.Ast.Soc ,*237*, pp341-354, 1988

Spectrum Management and the impact of the GLONASS and GPS satellite systems on radio-astronomy.
J.E.B. Ponsonby
Proceedings of the West German-Soviet Workshop on
"Scientific and Industrial collaboration on the use
of the Global navigation system GLONASS",
Leningrad, Sept 1990

Spectrum Management and the Impact of the GLONASS and GPS satellite systems on radio-astronomy.
J.E.B. Ponsonby
Journal of Navigation, Vol.44, 3, Sept 1991

Spread-Spectrum: What it is, what it is for, its properties, with recommendations for minimising its deleterious side-effects, particularly on radio-astronomy.

J.E.B. Ponsonby

This paper has been submitted to Radio Science

www.ingramcontent.com/pod-product-compliance
Lightning Source LLC
LaVergne TN
LVHW050302210926
848481LV00014BA/801